MW01485137

Simplified GD&T

(Based on ASME Y14.5-2009)

ASHOK KUMAR

Copyright © 2019 Ashok Kumar

All rights reserved.

ISBN: 1980948712
ISBN-13: 9781980948711

DEDICATION

Dedicated to my country, India, which I love most!

CONTENTS

ACKNOWLEDGMENTS

My parents, my roots, who are my timeless
source of support and encouragement.

My elder brother, Mr. Anil Kumar,
who taught me to dream big, and get it.

My better-half, Archana, walking my dreams
together, through all terrains of life, mostly
pleasant for me, tough for her.

My daughters, Ashna, Aadya, and Ananya,
providing cheer and continuous energy.

My technology inspirations from various industries,
who hooked my zeal to technical excellence
and I travelled, a bit, path of my heroes:
Prof. Narayan Rangaraj
Vivekanand Sawant
Saumitra Chattopadhyay
Anil Pagadala
John Clark
& more..

My team, whose belief in me makes me more
presentable than what I am ☺

1 BIRTH OF GD&T

It was World War II and the place was Great Britain. Everyone was working hard to fulfil all the requirements of their country. Manufacturing industry was also on its toes with the expectation of maximum production with minimum defects. There came a design for a 5 mm radius hole to be made on a plate. Location of the centre of the hole was (10, 5) with an acceptable positional tolerance of 1 mm. It was drawn in the coordinate system as shown below:

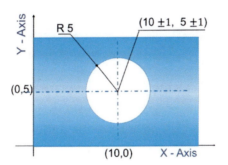

Figure 1.1 – Design of a hole with positional tolerance

The manufacturer converted above drawing into a manufacturing drawing to keep the centre in a grey shaded area represented by four corners (11,4), (11,6), (9,6), and (9,4) as shown below, in figure 1.2, as per conventions of coordinate system dimensioning and tolerancing.

1

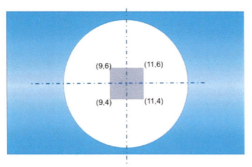

Figure 1.2 –Manufacturing interpretation of tolerance

When manufactured parts arrived then many parts, whose centre of the hole was near corners of the shaded area, were rejected because the centre of such holes was more than 1 mm away from the ideal position at (10,5). The designer intended the green circular area, as shown in figure 1.3, as an acceptable area for the positions of the centre of the hole.

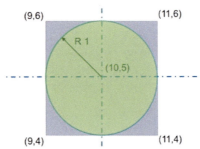

Figure 1.3 – Designer interpretation of tolerance zone

This gave rise to conflict between manufacturer and designer. The area of conflict lied between grey square area and green circle are as shown below, in figure 1.4, by red hatched area.

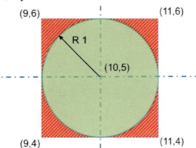

Figure 1.4 – Area of conflict (shaded with red color)

Neither did the designer agree to accept those parts falling in the area of conflict, as it would not meet the functional requirement of the part, nor did the manufacturer agree to accept the rejection, as it was compliant to the tolerances provided in the drawing.

Since it was war period where time was a crucial factor, this conflict pulled the attention of many people, including **Stanley Parker**, who came up with an idea of providing geometric positional tolerance which will be represented by the circular area around the ideal position.

This is how "**Geometric Dimensioning & Tolerancing**" (**GD&T**) took birth in the 1950s.

In addition to positional tolerance, there were few more issues identified, for example, tolerance in straightness, circularity, flatness, etc., which were all combined together to give it a shape of today's GD&T, as defined in ASME Y14.5-2009, which is the output of international collaboration.

2 DEFINING GD&T

GD&T stands for Geometric Dimensioning and Tolerancing.

Definition

GD&T is a mechanical engineering *language* to communicate a *functional requirement*.

Purpose of the book is to learn this language.

Explanation of above definition

Let's consider the example given in the previous chapter. The hole is to be put on tip of a rifle. There will be a pointer to be placed in the hole using which soldier will fire at aim. If the position of the hole is displaced, then the aim will be incorrect. As a result, the functional requirement of the rifle will not be met. Here the rifle's functional requirement determines the functional requirement of the hole.

Let me give you another example of a functional requirement. Suppose you are designing fuel injection nozzle for a diesel engine. You want the nozzle to be fixed at a particular position and angle such that injected fuel get properly mixed with air before combustion starts. You may provide positional and angular tolerances such that fuel is still mixed properly with air before combustion to make sure the desired functionality is achieved.

Need yet another example of functional requirement? Look at figure 2.1 in which a bracket is shown on the left side with the requirement of holes on it.

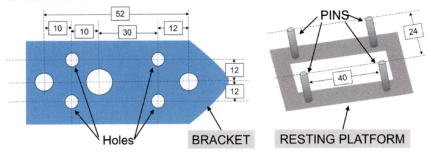

Figure 2.1 – Design of a bracket with holes

It needs to rest on the resting platform as shown on the right side in Figure 2.1 which has four guiding pins fixed to it. When the bracket is placed on the resting platform, it must fit properly. It is a functional requirement. Tolerances on pins and holes can be given only until this functionality is met.

There can be numerous such examples of functional requirements. The core concept remains the same, that is, within provided tolerance, the part or assembly should be functional as designed.

Before we dive deep in to GD&T, it is better to quickly glance through our concepts of dimensioning and tolerancing in the next two chapters.

3 DIMENSIONING

Traditionally, we have a comprehensive set of conventions for dimensioning. We provide tolerances with the help of "±", e.g., 10±1, as we saw in the case discussed in chapter 1. Let us look at different types of dimensioning.

§   **Linear dimensioning**   §

There are 5 types of linear dimensioning:

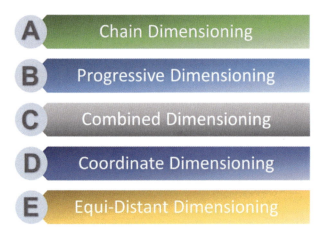

Look at figure 3.1, given below, which shows all five types of linear dimensioning.

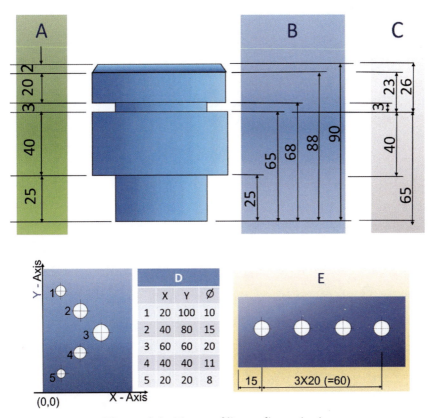

Figure 3.1 – Types of linear dimensioning

§ Circular, Spherical, and Square dimensioning §

Below are the symbols used for these dimensioning.

Symbol	Object type
Ø	Circular diameter
SØ	Spherical diameter
R	Circular radius
SR	Spherical radius
☐	Square side

Refer to figure 3.2 for examples of these dimensions.

7

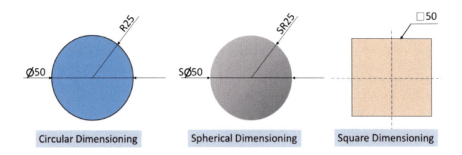

Figure 3.2 – Circular, spherical and square dimensioning

§ **Arc dimensioning** §

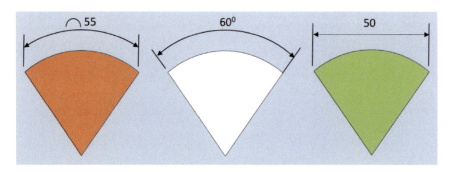

Figure 3.3 – Arc dimensioning

§ **Chamfer, and countersunk dimensioning** §

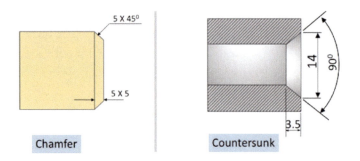

Figure 3.4 – Chamfer and countersunk dimensioning

§ Depth and alternative countersunk §

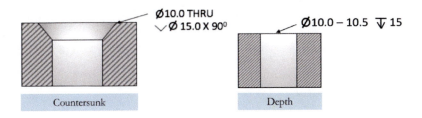

Figure 3.5 – Depth and alternative countersunk

§ Taper dimensioning §

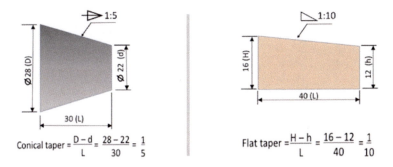

Figure 3.6 – Taper dimensioning

§ Counterbore and spotface §

The counterbore is a stepped hole with a uniform diameter. Spotface (SF) is the facing of a certain part, especially for cast or forged items. It is generally done around the hole and looks more like a sallow counterbore

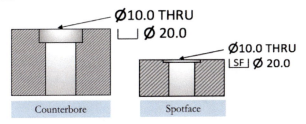

Figure 3.7 – Counterbore and spotface

§ Basic dimension §

A basic dimension is a theoretically ideal dimension, without any tolerance. Designs are based on basic dimensions and tolerances are provided on top of it. These are usually mentioned in a rectangle as you can find in Figure 2.1 given in chapter 2.

§ Reference dimension §

These are the dimensions provided for information purpose only, not for the manufacturing process. Generally, these are not provided with tolerance but if tolerance is needed for information purpose then they can be included. Before the use of CAD software, it was represented with notation REF as "**1500 mm REF**" however in CAD software it is generally represented with parenthesis as "**(1500 mm)**", as shown in figure 3.8, given below.

(1500 mm)	1500 mm REF
Used in CAD software	**Traditional method**

Figure 3.8 – Reference dimensioning

A reference dimension is generally driven by other values on drawing and does not govern production or inspection operations.

3.1 BASIC DIMENSIONING RULES

Let us look at basic dimensioning rules followed across the world, as given below:

- Each dimension should be associated with a tolerance (except for reference, max/min).
- Measuring dimension from drawing or assumption of a distance or size is not permitted.
- Provide only the necessary dimensions to complete the definition.
- Usage of reference dimension (one with a reference to another dimension) should be minimized.
- Especially for mating parts (hole and Shaft), dimensions should lead to only one interpretation.
- Manufacturing method (like drilled, punched) should not be given, the only dimension is to be provided. (Except when processing, quality, or environmental information is essential then specify on drawing or in a separate document)
- Wires, cables, sheets, rods, etc., which are made in gage or with code numbers, to be mentioned with diameter/thickness. Gage or code numbers can be put in parentheses next to dimension.
- All dimensions and tolerances are assumed at 20°C (68°F), else need to be mentioned.
- All dimensions and tolerances are assumed at free state condition (no stress).
- All tolerances apply to the full length, depth, and width of the feature unless mentioned.

In the next chapter, we will look at different aspects of tolerances, different terms associated with it and different cases of tolerance considerations.

You will come across many terms, for example, limit, allowance, tolerance, deviation, etc. which sound similar to each other but they all have a different meaning in engineering terms. I am sure you would enjoy it.
So why wait? Let's keep rolling!

4 TOLERANCES

Let's start with a basic question for you.

 Why we need tolerance?

Any manufacturing facility will have some inaccuracy when compared with intended dimensions. These inaccuracies may be in microns (1/1000 mm), but even that can be dangerous. Considering this limitation, Designers make designs with some allowed inaccuracies, called tolerance. Tolerance is given with the following considerations:

1. It is possible to manufacture part or component within given limits
2. The component will be able to function as intended
3. Assemblies are able to fit and work as per design

In the next section, we will learn about the fundamentals of tolerances and different terms associated with it.

4.1 FUNDAMENTALS OF TOLERANCES

Let us look at a shaft and try to understand the different fundamental terms of tolerancing

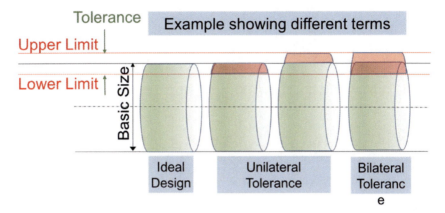

Figure 4.1 – Fundamentals terms used in tolerancing

Basic/Nominal Size *(Standard/ideal size)*
It is the standard or ideal size of a part, as shown by the leftmost shaft in figure 4.1. We apply deviation and tolerance with reference to the basic/nominal size. These are usually mentioned in a rectangle as you can find in figure 2.1 given in chapter 2.

Limits *(Upper and Lower limits of a dimension)*
These are maximum and minimum dimensions of an acceptable manufactured part. These limits are shown by two red lines in figure 4.1.

Tolerance *(Permissible variation in dimension)*
It represents the margin of variation, which is called tolerance zone. It is the difference between upper and lower limits, as shown in figure 4.1. It is decided based on the allowed inaccuracies without causing any functional trouble.

Unilateral Tolerance:
The dimension may vary only in one direction. Second and third shafts show unilateral tolerances.

Bilateral Tolerance:
The dimension of the part may vary in any direction as shown by the rightmost picture in figure 4.1.

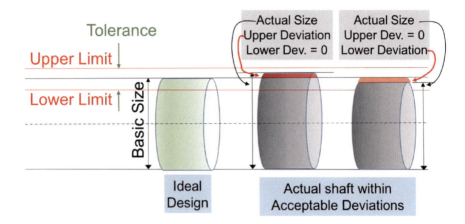

Figure 4.2 – Analysis of actual shaft tolerance

Actual Size:

It is the dimension of the actual part after manufacturing. We can find two examples in figure 4.2.

Deviation

It is the difference between actual size and basic size.

Upper Deviation: Difference between the <u>maximum</u> actual size and the corresponding basic size. It is first case (left grey shaft), in which the actual size is slightly bigger than basic size which causes upper deviation only (lower deviation is zero).

Lower Deviation: Difference between the <u>minimum</u> actual size and the corresponding basic size. It is the second case (right grey shaft), in which the actual size is slightly smaller than basic size which causes lower deviation only (upper deviation is zero)

 What is relationhip between tolerance and deviation?

1. *"Tolerance is the 'allowed limits' of deviations, of a dimension, from its basic/nominal value".*
2. *"Tolerance is unintentional/unwanted but allowed deviation".*

4.2 FITS AND DEVIATIONS

In the previous section, we saw tolerance of an independent part or component. In this section, we will learn about the tolerances of mating parts, means two interacting parts. Consider a case of piston and cylinder in which we want loose fitting between them. We would define tolerances of piston and cylinders in such a manner that for the entire range of combined tolerances, there remains a gap between them such that loose-fitting is achieved and motion is possible.

In order to better understand the concepts of interacting parts, we will study typical concepts of fits. Fits are studied as a mating mechanism between a shaft and a hole. There are three types of fits as shown in figure 4.3:

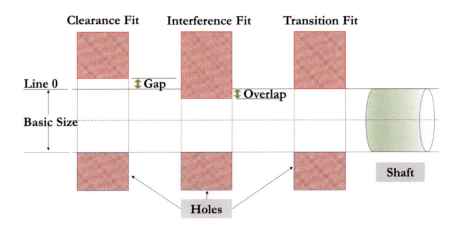

Figure 4.3 – Types of fits

There are three types of fits:
1. **Clearance Fit:** When the shaft is always smaller than the hole. It is used for free rotation.
2. **Interference Fit**: When rotation or relative motion is not allowed. In this fit, the shaft is always larger than the hole.
3. **Transition Fit:** Depending on actual sizes, it can lead either to clearance or interference fit.

Design of clearance fit

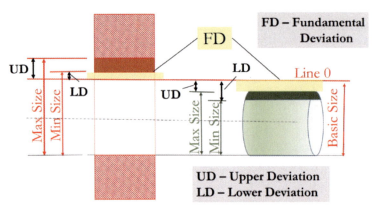

Figure 4.4 – Design of clearance fit with tolerance

Take some time, look at Figure 4.4, given above, and then read below statements carefully:

- You always have a **Line 0** (zero) at the basic size.
- Depending on the type of fit needed, you will define gap or overlap.
- For the above clearance fit case, we will define minimum gaps between hole and shaft with reference to line 0. These are shown by yellow strips in figure 4.4. These are called **"fundamental deviations (FD)**.
- FD gives you the minimum size of the hole and maximum size of the shaft to ensure the minimum desired gap between them for clearance fit.
- FD for hole gives you lower deviation and FD for shaft gives you upper deviation.
- Now adding tolerances. Red strip is tolerance given to hole. It allows you to make the hole bigger than the minimum permissible size. Green strip is tolerance given to shaft. It allows you to make it smaller than maximum permissible size.
- Tolerance of hole decides upper deviation (UD) of the hole which is the sum of FD and tolerance.
- Tolerance of shaft decides lower deviation (LD) of the shaft which is again the sum of FD and tolerance.

? Question time !

Question: Why fundamental deviation is given such a name? Does it define any fundamental property of fit?

Answer: Yes, it defines the type of fit. The fundamental deviation is an intentional deviation from the basic size to achieve the intended functionality. For example, the hole will be made bigger and shaft will make smaller to create a gap between hole and shaft for a clearance fit.

Question: Find out correct statements for clearance fits:
 A. For holes, UD = LD + Tolerance
 B. For holes, UD = FD + Tolerance
 C. For shafts, LD = UD + Tolerance
 D. For shafts, LD = FD + Tolerance

Answer: If you find all statements to be correct then you are good to proceed, else you must revise the content up to this point and clarify the doubt before proceeding further.

Question: What are a basic shaft and basic hole? *(Hint: the answer can be derived based on terms defined earlier.)*

Answer: These are shafts and holes having zero fundamental deviations, that is their sizes are the same as the basic size of the fit. Obviously, the basic hole will have zero lower deviation and basic shaft will have zero upper deviation.

Understanding Allowance

The allowance is minimum intended/planned gap between mating parts. It is decided on the basis of:
 1. Type of fit needed
 2. Tolerance needed due to manufacturing limitation.

Consider a case of 10 mm basic size clearance fit. It is known that in reality, the grinding operation that produces the final diameter may introduce a certain small-but-unavoidable amount of random error. Therefore, the engineer specifies a tolerance of ±0.01 mm ("plus-or-minus" 0.01 mm) such that actual shaft would be available between 9.99 to 10.01 mm. If we apply the same logic for the hole then actual hole size may also be expected between 9.99 to 10.01. Some shafts will fit in some holes when their diameter is smaller than the diameter of a hole, but others will not fit. In order to ensure clearance fit for all parts, the designer will have to increase the size of the hole or reduce the size of the shaft. Designer keeps the size of shaft same and increases the size of the hole to 10.03 mm with a tolerance of ±0.01 mm. In this case, hole dimension will vary between 10.02 mm to 10.04 mm. With this design, the **minimum gap** between smallest hole (10.02 mm) and largest shaft (10.01 mm) will be 10.02-10.01 = 0.01 mm.

Let's take it one step ahead. The designer wants the minimum gap to be 0.02 mm for consideration of vibration in the machine, he will do the reserve calculation and increase the hole size to 10.04±0.01 mm which will result in a minimum intended gap of 0.02 mm.

The intended minimum gap is called ALLOWANCE.

In Figure 4.5, given below, you can see allowance highlighted on the right side.

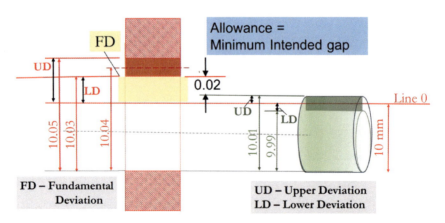

Figure 4.5 – Allowance in a clearance fit

Systems of Fits:
Hole basis systems and shaft basis systems

Did you notice one important aspect in the above example? The designer kept the shaft dimension unchanged and increased hole dimension. They had two more solutions:

1. Keep hole dimensions unchanged and reduce shaft dimension to design for 0.02 mm allowance.
2. Increase hole dimension and reduce shaft dimension to achieve 0.02 mm allowance.

In order to control the cost of operations, generally, we keep dimension of one component as fixed and manipulate dimension of another component.

*   If we keep the dimension of the hole as fixed, then it is known as **Hole Basis System**. Here the minimum hole diameter is kept the same as the desired basic size of the assembly.
*   If we keep the dimension of the shaft as fixed then it is known as **Shaft Basis System**, in which the maximum diameter of the shaft is kept at a desired basic size of the assembly.

Look at Figures 4.6 and 4.7, given below, for a visual explanation.

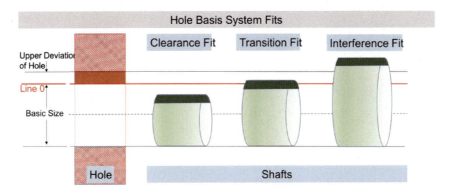

Figure 4.6 – Illustration of the Hole Basis System

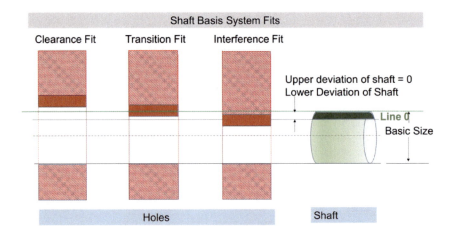

Figure 4.7– Illustration of the Shaft Basis System

Selection of hole or shaft basis system

Question: Which is a preferred system, hole basis and shaft basis, why?

Answer: Hole Basis System is preferred over Shaft Basis System. This is primarily due to the following two reasons:

1. Holes are made from available drill bits which comes in pre-defined incremental sizes. Therefore, we choose a hole of available drill bit size and then plan shaft size as per functional need.

2. It is easier to change the shape of the shaft by machining on a lathe. It is a single operation process. On the other hand, creating a hole of arbitrary size will be a two-step process, the first drill with nearest smaller size drill bit and then bore on a lathe to the desired size.

Therefore, usually, the Hole Basis System is used across the world.

❉ **Your thought bite** ❉
There may be a functional need to choose Shaft Basis System. Can you think of any such case?

Hint: Think about the turbine propeller shaft, which is a critical component.

20

Extention of system of fits

The system of fits is described with the help of a hole and shaft combination but it is not limited to only holes and shafts. **The concept of hole and shaft is generic in nature.** Hole and shafts are used to designate all the external shape (for shafts) and internal shape (for holes), not necessarily cylindrical. Therefore, it extends to all curved types of fits.

Example 1: Piston and cylinder in an engine.

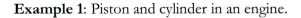

Figure 4.8 – Sample of piston and cylinder of a bike

In this example, you consider piston as the shaft and cylinder as the hole. Definitely, you need clearance system of fit. Right?

❋ **Your thought bite** ❋
There are two prominent types of motion between hole and shaft for clearance fit case. Can you guess what are those?

Hint: Translational and rotational.

Example 2: Crankshaft and connecting rod

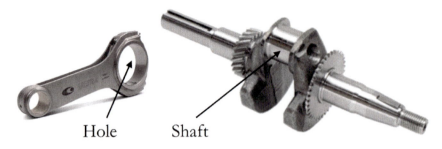

Hole   Shaft

Figure 4.9 – Sample of connecting rod and crankshaft

In this example, the internal surface of the connecting rod is treated as a hole and the upper surface of crankshaft bearing is considered as the shaft.

Example 2: Hole and shaft in a gear assembly

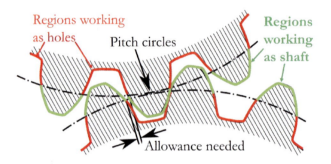

Figure 4.10 – Hole and shaft profiles in gear assembly

Here you can see the outer surface of addendum (highlighted with green colour) behaves as a shaft and outer surface of dedendum (highlighted with red colour) behaves as a hole.

❈ **Your thought bite** ❈
Can you identify hole and shaft profiles in camshaft assembly?

4.3 ANGULAR TOLERANCES

For angular dimensions, we may provide some angular tolerances to accommodate manufacturing limitations to produce parts with ideal angles. It is shown below in figure 4.11 where angle may vary by 1 degree on either side.

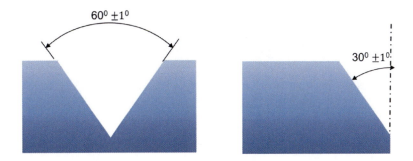

Figure 4.11 – Illustration of angular tolerance

Here the deviations are in terms of angle, not in length. This is a major difference between linear tolerance and angular tolerance.

In this chapter, we learnt about:
- Why we need tolerance
- Basic/Nominal size
- Upper and lower limits
- Tolerance - unilateral and bilateral
- Upper and lower deviations
- Types of fits - clearance, transition, and interference
- fundamental deviation
- Allowance
- Systems of fits - hole basis and shaft basis
- Extension of the system of fits
- angular tolerance

If any topic is unclear, then read it again before proceeding.

5 FUNDAMENTALS OF FEATURES

GD&T is applied to any **FEATURE** to provide details of geometric tolerances. Before we understand geometric tolerance, let's first understand the meaning of a feature.

5.1 WHAT IS A FEATURE?

In mechanical engineering drawing or CAD, the feature is used for cuts, protrusions, rounds, fillets, etc. In fact, it represents almost all physical portion (curved or flat) of a part. It may include different types of holes, counterbore, countersink, counter drill, taper hole, step bore, slots, chamfer, pockets, bosses, and grooves. Look at figure 5.1 for a sample part with few features.

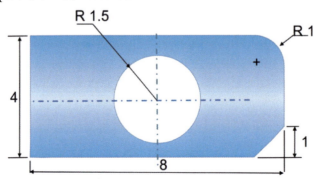

Figure 5.1 – Part with multiple features

There are nine features in the above diagram. Can you identify them? Let me count for you. 4 sides + 1 fillet + 1 chamfer + 1 cylindrical hole + 1 top surface + 1 bottom surface = 9 features.

Look at figure 5.2 given below to be familiar with a few real-life samples of features. There are uncountable types of features and you can also create your own based on your design needs.

Figure 5.2 – Real life feature samples

Let us add a few more features on the same plate as shown in figure 5.3 below. I want to measure the dimension of all features by using either external jaws or internal jaws of Vernier calliper by "**firmly**" positioning them on "**opposite points**" of a feature. I can measure feature A, C, the long and short axis of D, upper & lower gaps of E, and thickness of the plate. I cannot measure the dimension of B, F, and G, because they do not have mutually *__opposite__* resting place for both arms of jaws.

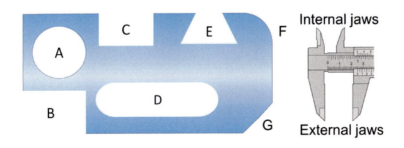

Figure 5.3 – Examples of features to identify FOS

5.2 WHAT IS FEATURE OF SIZE?

In the above example, there are few features dimensions of which can be measured by means of **two opposite physical points**, are a special type of features, called **Feature of Size (FOS)**. These are very useful features because you identify their point of centre (for circular profile), the central axis (for cylindrical holes), central plane (for parallel surfaces) and use them as a reference for any purpose. So you can say, FOS is a special feature which generates a point of reference with help of two opposed physical points available on them.

To summarize, we can say, FOS is
- any circular, cylindrical, spherical, or parallel set of surface
- with available two opposed physical points
- with given tolerance dimension.

All three conditions are mandatory to derive point, axis or plane of reference, so these are requirements of FOS.

The feature of size dimensions is of two types: an **internal feature of size and external feature of the size dimension**. In the above drawing, the centre hole is an internal feature of size dimension and the thickness of the plate is an external feature of size dimensions.

ASME 14.5-2009 defines same FOS as **a regular feature of size** as:
"One cylindrical or spherical surface, a circular element, and a set of two opposed parallel elements or opposed parallel surfaces, each of which is associated with a directly toleranced dimension"

There is another type of FOS which is called **Irregular Feature of Sizes**. Below are definitions as given in ASME Y14.5-2009.

1.3.32.2 Irregular Feature of Size: the two types of irregular features of size are as follows:
a) a directly toleranced feature or collection of features that may contain or be contained by an actual mating envelope that is a sphere, cylinder, or pair of parallel planes

b) a directly toleranced feature or collection of features that may contain or be contained by an actual mating envelope other than a sphere, cylinder, or pair of parallel planes

In simple terms, irregular FOS are those FOS which can be hypothetically enclosed by any other shape to derive reference point, axis or plane based on enclosing shape. Type (a) are those irregular FOS which can be hypothetically enclosed in a sphere, cylinder, or pair of parallel plains. Type (b) are those enclosed by other than sphere, cylinder, or pair of parallel plains.

Refer to Figure 5.4 for type (a) irregular feature of sizes. It shows the derived reference plane of reference obtained from a hypothetical enclosure for irregular FOS.

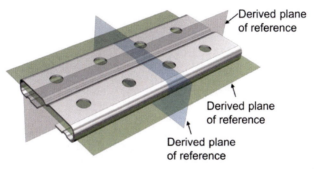

Figure 5.4 – Type (a) Irregular feature of size

Refer to Figure 5.5 for type (b) irregular feature of sizes. It shows derived axis, planes and hypothetical cylinder reference obtained for irregular FOS.

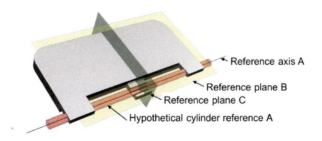

Figure 5.5 – Type (b) Irregular feature of size

Non-feature of size: These are features with dimensions but do not qualify as FOS. In our example, chamfer, fillet, etc. fall into this category.

Feature & FOS 🧢 recap

- A feature is any shape, face, cuts, protrusions, rounds, fillets, etc. to create any part shape in the mechanical industry.
- Regular Feature of size (FOS) is any circular, cylindrical, or parallel set of surfaces with dimension and tolerance and having two opposite point physically available on the feature to measure, manufacture and establish the point, axis or plane references.
- An irregular feature of size is not having shapes given above. If it can be hypothetically enclosed in any regular shape (sphere, cylinder, set of parallel planes) then it is called type (a), but it is enclosed by other hypothetical shapes then type (b) irregular feature of sizes.
- Non-features of sizes are those feature which has dimensions and tolerances but cannot fulfil all criteria, for example, opposite points are not available on the feature, are called non-feature of sizes.

A feature of size is used extensively in GD&T to establish lots of references, we will see as we are going to learn about further topics.

5.3 FEATURE MATERIAL CONDITIONS

Due to manufacturing tolerances, any feature will have an upper limit and lower limit of size. Accordingly, the amount of material will be maximum or least. The maximum material of shaft results in the largest dimension and least material of shaft results in the smallest dimension. For holes, it is exactly opposite. Maximum material remaining after cutting hole results in smallest dimension and least material of after cutting hole results in largest hole. These material conditions of features are called feature material condition and termed as:

1. MMC – Maximum material condition
2. LMC – Least material condition

Look at illustration given below to confirm your understanding.

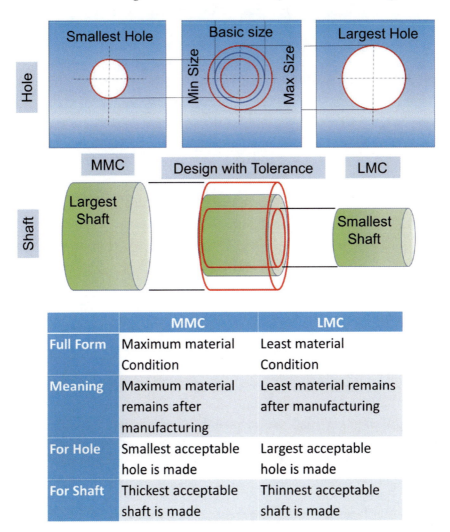

Figure 5.6 – Illustration of MMC and LMC

	MMC	LMC
Full Form	Maximum material Condition	Least material Condition
Meaning	Maximum material remains after manufacturing	Least material remains after manufacturing
For Hole	Smallest acceptable hole is made	Largest acceptable hole is made
For Shaft	Thickest acceptable shaft is made	Thinnest acceptable shaft is made

I would like to put a quick question to check your understanding at this level. Consider the design of the clearance fit we discussed in the previous chapter. We learnt about the minimum and maximum size of hole and shaft and then we learnt about allowance. Can you re-present the same allowance in terms of MMC/LMC of hole and shaft? Try it before reading further.

Answer:
Allowance in clearance fit = minimum gap between hole and shaft

The minimum gap will be a situation when the hole is smallest and the shaft is largest, means, the hole is at MMC and shaft is also at MMC. It implies:

Allowance = MMC hole – MMC shaft.

Now attempt to re-present the following values in terms of MMC/LMC:
1. Maximum clearance
2. Minimum interference
3. Maximum interference

You should celebrate the day if your answers are:
1. LMC hole – LMC shaft
2. LMC shaft – LMC hole
3. MMC shaft – MMC hole

5.4 TOLERANCE WITH MATERIAL CONDITION

For any assembly or Fits, it is important to consider the material condition of both parts (hole and shaft) to ensure proper functioning. Let's take two examples to understand the importance:

§ **Case 1: Clearance fit analysis with material condition** §

For a clearance fit, biggest shaft diameter should be smaller than the smallest hole diameter and there must exist a minimum gap (allowance) to ensure proper clearance fit functioning. We learnt earlier:
Biggest diameter shaft = MMC of the shaft -- Least desirable condition
Smallest diameter hole = MMC of the hole -- Least desirable condition

Therefore, our functional requirement should be:
MMC of Shaft < MMC of Hole
It guarantees clearance fit even at *least desirable conditions.*

? Question time !

Does the condition mentioned above (MMC of Shaft < MMC of Hole) always guarantees a clearance fit?
The answer is NO. Let's try to understand the situation.

You got a task to **design a hole** to meet the following requirements:
- Clearance fit with allowance (minimum gap) of 1 mm
- The shaft of MMC diameter of 8 mm. It cannot change.
- Dimensional tolerance of hole is ±1 mm

Design step 1: Calculate MMC of hole
= MMC of shaft + allowance = 8 + 1 = 9 mm.
Design step 2: Calculate nominal diameter of hole
= MMC of hole + tolerance = 9 + 1 = 10 mm
Design step 3: Calculate LMC of hole
= Nominal hole diameter + tolerance = 10 + 1 = 11 mm

Figure 5.7, given below, illustrates the above design.

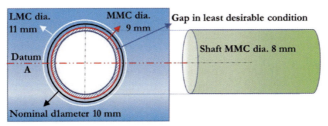

Figure 5.7 – Example of clearance fit without geometric tolerance

Here you see a new word "DATUM", which is nothing more than another name of reference. It can be a point of reference or line of reference or plane of reference. We will learn, in detail, about it in next chapter.

In the above example, we assumed the shaft axis to be perfectly aligned to the axis of the hole (datum A), which may not always be possible. Suppose there exists a manufacturing limitation in which maximum possible accuracy is limited to only 1 mm, means the position of shaft axis may deviate up to 1 mm in either direction from datum A (which is hole's axis). The zone in which shaft axis may be expected is shown below in figure 5.8.

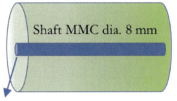

Figure 5.8 – Positional tolerance zone for shaft axis

Consider a case when the shaft in at lowermost position resulting in the situation illustrated in figure 5.9, given below.

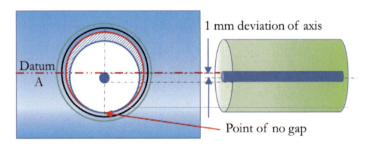

Figure 5.9 –Effect of positional deviation of shaft

Here you can see the shaft touching at the bottom of the hole. It can occur at any point on the hole. This situation is undesirable as one of the design requirement is to ensure 1 mm allowance. So our design failed. To fix this issue, we need to add positional tolerance in our earlier approach. Let's see how it goes:

Design step 1: Calculate MMC of the hole
= MMC of shaft + allowance + positional tolerance
= 8 + 1 + 1 = 10 mm.
Design step 2: Calculate nominal diameter of hole
= MMC of hole + dimensional tolerance of hole + positional tolerance
= 9 + 1 +1 = 11 mm
Design step 3: Calculate LMC of hole
= Nominal hole size + dimensional tolerance + positional tolerance
= 10 + 1 + 1 = 12 mm

Final design after considering dimensional tolerance, positional tolerance, MMC of the shaft and required allowance can be shown as in Figure 5.10 given below.

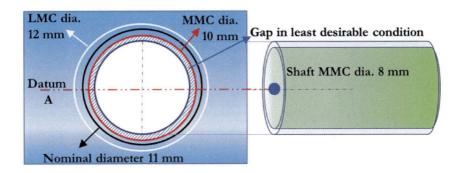

Figure 5.10 –Design with consideration of positional tolerance

Congratulations! You just designed a clearance fit while considering geometric tolerance, dimensional tolerance, allowance and MMC material condition. We are doing good!

Do you think we need to do the same analysis with LMC material condition or not? Why?

If your answer was "LMC analysis is not required, because we already did an analysis of the least desirable condition of MMC" then you are correct. Any deviation between MMC and LMC will be better than MMC, which is the worst-case fit.

§   **Case 2: Interference fit analysis with material condition**   §
For an interference fit, the shaft must have a close fit with a hole without any gap. Here the smallest shaft would be bigger than the largest hole as we learnt earlier:
Smallest diameter shaft = LMC of the shaft -- Least desirable condition
Biggest diameter hole = LMC of the hole -- Least desirable condition

Therefore, our functional requirement should be:
  LMC of Shaft > LMC of Hole
It guarantees interference fit even at *least desirable conditions.*

 Question time !

Does the condition mentioned above (LMC of Shaft > LMC of Hole) always guarantee a clearance fit?

Based on the previous case of clearance fit, you can guess the answer would again be NO. Right?

Wrong! Here additional positional tolerance will still keep the fit as interference fit due to the boundary of the hole. I leave it for you to draw your sketch to find out how it will show up?

Important point

With the help of the above two cases, we found MMC is to be considered while designing tolerance. On the other hand, LMC is not so important when designing for tolerances.

Then a question arises when would we use LMC?

One of the most common use of tolerancing at LMC is to take care of any hole near the edge of a part, as shown in picture 5.11, shown below.

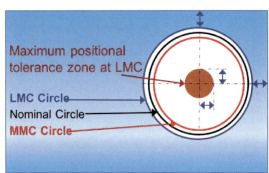

Figure 5.11 – LMC consideration for positional tolerance

Here positional tolerance should be considered at LMC (largest hole) such that hole to safely away from edge and part does not become weak. The maroon round area shows the maximum positional tolerance zone at LMC to avoid hole touching the edge.

5.5 TOLERANCE WITHOUT MATERIAL CONDITION

There are few geometric tolerances which are independent of feature size. For example, if any surface is to be flat or circular then the size of the feature does not matter. Such material condition is called **RFS**, that is, regardless of feature size. We will take it up later when we do a detailed analysis.

In this chapter, we learnt about:
- Features
- Features of size
  - o   Regular feature of size
  - o   Irregular feature of size (type a and b)
  - o   Non-feature of size
- Feature material conditions
  - o   MMC (Maximum material condition)
  - o   LMC (Least material condition)
- Tolerance with material conditions
- Tolerance without material conditions
- Material conditions modifiers

Clarify all doubts before proceeding to the next chapter.

6 DATUM – GD&T SYSTEM OF REFERENCE

In this chapter, we will learn about all types of datum used extensively in GD&T to measure dimensions and tolerances.

6.1 DEFINING DATUM

The concept of Datum is used in GD&T to provide a reference for dimensions and tolerances. It is a Latin word. meaning "given", i.e. an accepted fact. Let's understand Datum and related terms.

§ Datum reference frame §

Suppose you are given an engineering drawing with dimensions and tolerances to perform some machining activities to produce a part. How would you proceed? There are two primary needs:

1. **Restricting** movement of the part on some clamp, or so, to perform any operations. In engineering terms, it is a restriction of the **degree of freedom.** (Recall your studies of six degrees of freedom for a free body in which three are translational and three are rotational movement along X, Y and Z axes, this is also shown in Figure 6.1).

2. Next, you would need references like point, axis or plane from which dimensions can be measured. These references may be on any feature or we may create an imaginary reference from which dimensions are taken or controlled.

Datum reference frame is created to fulfil the above two requirements. Refer to Figure 6.1 and notice planes A, B, and C forming the frame in which the object is placed near the origin.

In a practical scenario, to fix any part on three planes, you would go in a sequence. Let's depict the same process while fixing the object shown in Figure 6.1 to the three planes A, B, and C. Let us first put plane 'A' (coffee color plane at the bottom) to establish full contact with the object. Plane 'A' will restrict the movement of an object in the Z direction, although the object can still move in X and Y direction. Plane 'A' will also restrict the object's rotation around X and Y direction (else full contact between the object and plane 'A' will be lost) but it can rotate around Z direction while keeping full contact. Look at the right side table in Figure 6.1 where plane (datum) A restricts three degrees of freedoms as explained.

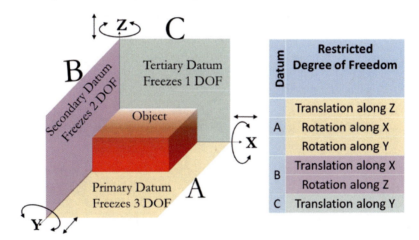

Datum	Restricted Degree of Freedom
A	Translation along Z
	Rotation along X
	Rotation along Y
B	Translation along X
	Rotation along Z
C	Translation along Y

Figure 6.1 – Datum reference frame

Now use plane (datum) 'B' to further restrict the movements of the object. It will restrict the translational degree of freedom of the object along the X-axis and also a rotational degree of freedom along Z axis to keep the object in full contact with a plan (datum) 'B', as shown in the table on the right side.

As the last step, use a plane (datum) 'C' to further restrict the movements of the object. It will restrict the translational degree of freedom of the object along Y-axis to keep the object in full contact with a plan (datum) 'C'.

With the help of the three planes (datums) A, B, and C, all degrees of freedoms are restricted and the object will not move at all. These three planes or datums will also become a reference for dimensioning and tolerancing.

This entire setup is called the datum feature frame.

Did you notice the terms primary, secondary and tertiary datum in the figure above? Below is the convention to call them so:
- **Primary Datum** restricts 3 degrees of freedom
- **Secondary Datum** restricts 2 degrees of freedom
- **Tertiary Datum** restricts 1 and last degree of freedom

We interchangeably used plane and datum in the above explanation. It was an intention to start imbibing the term datum in your mind. Now let us define a few concepts related to the datum feature frame.
- **Datum:** These are "**imaginary**" reference point, surface, or axis on an object against which measurements are made. In the above example, plane A, B, and C were datums, as we started using the term for those planes.

- **Datum feature:** "**Actual**" reference point, surface, or axis on an object feature. Example, the face of an object.

- **Functional Datum:** Datums are chosen based on the connection between parts to fulfil its functionalities. Functional datums are not used for manufacturing.

- **Manufacturing Datums:** These are datums for the manufacturing process to save cost, improve process speed, and repeatability. Tolerance analysis is used to get manufacturing datums from functional datums. Example,
  Functional tolerance + Bonus Tolerance = Manufacturing tolerance).
  We will learn about bonus tolerance later in this book. The above equation is also given here to differentiate between functional and manufacturing tolerances.

- **Datum Feature Simulator (theoretical):** It is the theoretically perfect boundary used to establish a datum from a specified datum feature, as shown for a shaft and hole in figure 6.2, as given below:

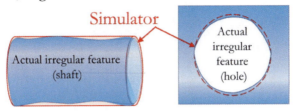

Figure 6.2 – Datum feature simulator (theoretical)

- **Datum Feature Simulator (Physical):** It is the physical boundary used to establish a **simulated datum** from specified datum features. Look at Figure 6.3 in which Step 1 is showing the placement of an object one a primary datum simulator to establish a primary datum. This step restricts three degrees of freedom as explained with the help of Figure 6.1.

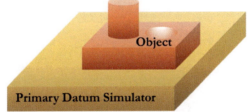

Figure 6.3a: Step 1: Primary datum simulator

In the next step, another plate (secondary datum simulator) is placed on one side of the object to establish a secondary datum. This step restricts two more degrees of freedom.

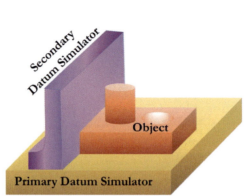

Figure 6.3b: Step 2: Secondary datum simulator

In the last step, another plate (tertiary datum simulator) is placed on one side of the object to establish a tertiary datum. This step restricts one, which is the last degree of freedom.

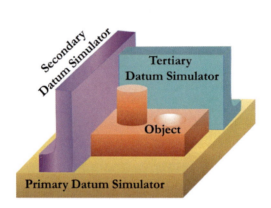

Figure 6.3c: Step3: –Tertiary datum simulator

§   **Datum Symbol – Representation of Datum**   §

After identifying any datum, we need to identify it in drawing too. Below are the conventions followed to show a datum.

Datum Symbol

Figure 6.4 – Datum symbol

As we learnt, the datum may be a point, line or surface, below are the example to represent them in all these cases.

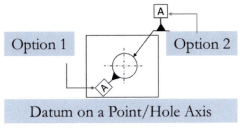

Figure 6.5 – Point or hole axis as Datum

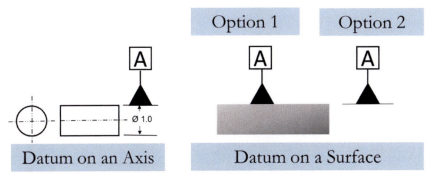

Figure 6.6 – Axis and surface as datum

6.2 DATUM TARGET IN GD&T

Because of irregularities and large size, entire feature surface cannot be effectively used as a datum feature, for example, uneven casting, forging, thin surface subjected to bending, warping, etc. To overcome such situations, "**Datum Target**" is used, which is a placeholder on Datum feature to derive Datum. It may be a "**point**", "**line**" or "**plane**".

Going to the basics of geometry, at least three non-linear points are required to define a plane. These points can be located with help of 3 pins, either with a pointed tip or flat top or spherical top, as shown in figure 6.7, given below.

Pointed Pin   **Flat top Pin**   **Round top Pin**

Figure 6.7 – Types of pins

Let's consider a case of a large uneven casting, shown in Figure 6.8, given below. We want to determine the plane of the casting to make it Datum feature. Practically, how would we do it? To mark three places on casting surface, let's select 3 flat top pins and mark target touching area as shown by a blue hatched circle. Using the location of three pins, we can determine the Datum plane.

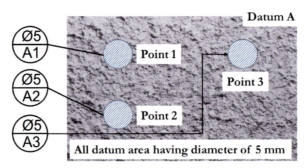

Figure 6.8 – Identification of Datum target

Datum target is written in a circle. Lower semicircle has Datum name and target area number. Upper semi-circle has the size of datum area.

There can be two deviations in the sample given above.
    (i)   For point target areas, the upper circle will be blank.
    (ii)   If dimension cannot fit in an upper semicircle, then write it outside.

Look at Figure 6.9, given below, for examples of three hypothetical points A4, A5, and A6 to explain above two situations.

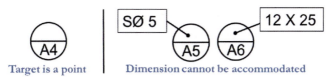

Figure 6.9 – Identification of Datum target

Important notes:

1. With a similar approach, we use two pointed pins to determine datum line (as only 2 points are required to draw a line), and, 1 pointed pin to determine datum point.
2. Datum identification tool need not be only a pin. It can be any other tool with a different contact area, like rectangular, square, etc. The only consideration is it should be convenient to use.

In this chapter, we learnt about:

- Datum – a system of reference in GD&T
- Datum feature
- Datum feature frame
    - o Primary, secondary and tertiary datums
    - o Relationship between datums and degree of freedoms
- Functional datum
- Manufacturing datum
- Datum feature simulator (theoretical/physical)
- Showing datum symbol on point, axis, and surface
- Datum target and its representation

If you are unclear about any concept, then first clear it before proceeding to the next chapter.

7 GD&T LANGUAGE – SYMBOLS AND SYNTAX

In this chapter we will about basics of GD&T language, in particular:
- Symbols of fourteen (14) geometric tolerances
- Syntax to write geometric tolerance as per GD&T syntax
- The proper method of reading GD&T

Note 1: We will start with knowing the symbols of 14 geometric tolerances to first become familiar with language such that we can start writing GD&T language. Detailed analysis of these geometric tolerances will be covered in the next chapter. So, hold your curiosity till next chapter as you will have a little less clarity and lots of unanswered questions in this chapter, which will be made clear in next chapter.

Note 2: There are more symbols in addition to 14 geometric tolerance symbols. These are called modifier, for example, MMC, LMC modifiers. These modifiers are used to add a description to dimension or geometric tolerance to communicate more details. We will learn about the most widely used feature material condition modifiers in this chapter such that writing few examples are not impacted. Remaining modifiers will be taken up in next to next chapters.

7.1 SYMBOLS OF GEOMETRIC TOLERANCES

Refer to Figure 7.1, given below, which lists all fourteen types of geometric tolerances, their tolerance type, and application of these tolerances for independent or assembly features.

#	Geometric Tolerance	Tolerance Type	Application
1	Straightness	Form	Independent Features
2	Flatness		
3	Circularity		
4	Cylindricity		
5	Line Profile	Profile	Independent or Assembly
6	Surface Profile		
7	Angularity	Orientation	Assembly Features
8	Perpendicularity		
9	Parallelism		
10	True Position	Location	
11	Concentricity		
12	Symmetry		
13	Circular Runout	Runout	
14	Total Runout		

Figure 7.1 – 14 GD&T geometric tolerances

§   Tolerance type – Form   §

These are tolerances given to basic geometric shapes or FORMS. Refer figure 7.2, given below, listing straightness, flatness, circularity, and cylindricity as tolerances of type 'FORMS'. These tolerances are used to mention the kind of perfection needed for features with these shapes. All these tolerances are applied independently on the features. It means **datums** are **not required** for them as these features are controlled independently.

#	Geometric Tolerance	Symbol	Tolerance Type	Datum required?
1	Straightness	—		
2	Flatness	▱	Form	Datum is not required because features are measured independently
3	Circularity	○		
4	Cylindricity	⌭		

Figure 7.2 – Form geometric tolerances

§   **Tolerance type – Profile**   §

These are tolerances used for geometric curved PROFILE as given below in figure 7.3. These are related to curved surfaces. Line profile is the path of a 2D line and surface profile is a 3D profile of a curved surface. All these tolerances can be applied independently or with reference to a datum, which means datum may or may not be required, means **datum** becomes **optional** to mention.

#	Geometric Tolerance	Symbol	Tolerance Type	Datum required?
5	Line Profile	⌒	Profile	Datum may or may not be needed (optional)
6	Surface Profile	⌓		

Figure 7.3 – Profile geometric tolerances

§ Tolerance type – Orientation §

These are tolerances used for geometric ORIENTATION of a plane with respect to another plane (datum) as given below in figure 7.4. So datum becomes mandatory.

#	Geometric Tolerance	Symbol	Tolerance Type	Datum required?
7	Angularity	∠		
8	Perpendicularity	⊥	Orientation	Datum is mandatory to mention
9	Parallelism	//		

Figure 7.4 – Orientation geometric tolerances

§ Tolerance type – Location §

These are tolerances used for geometric LOCATION of a point, line or plane as given below in figure 7.5. Since all these tolerances are applied in relation to other feature, datum becomes mandatory.

#	Geometric Tolerance	Symbol	Tolerance Type	Datum required?
10	True Position	⊕		
11	Concentricity	◎	Location	Datum is mandatory
12	Symmetry	═		

Figure 7.5 – Location geometric tolerances

§ Tolerance type – Runout §

These are tolerances used for geometric RUNOUT of rotating parts. Runout is basically to highlight the deviation of rotating shafts from the desired mean stable position. Refer or Figure 7.6 for names and symbols. Since all these tolerances are applied in relation to other feature, datum becomes mandatory.

#	Geometric Tolerance	Symbol	Tolerance Type	Datum required?
13	Circular Runout	➚	Runout	Datum is mandatory
14	Total Runout	⫽➚		

Figure 7.6 – Runout geometric tolerances

7.2 MATERIAL CONDITION MODIFIERS

Material condition modifiers are modifiers of the tolerance to be provided to consider tolerance in consideration with material conditions. There are many more modifiers we will learn as we go along.

Below is a complete family of material condition modifiers to geometric tolerance and their symbols used in GD&T, as shown in Figure 7.7:

Figure 7.7 – Material condition modifiers

Yes, you saw it correctly. RFS has no sign. It means when you want to provide geometric tolerance independent of MMC and LMC, means without considering any material condition then you don't need to mention any modifier.

7.3 WRITING IN GD&T LANGUAGE - FEATURE CONTROL FRAME

As a designer, you need to communicate functional requirements. For each *Feature*, geometric tolerance *Control* is written in a *Frame*, which is called – "*Feature Control Frame*", as shown in Figure 7.8:

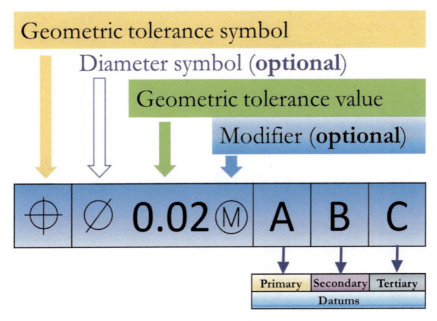

Figure 7.8 – Feature control frame

Notice the following details of feature control frame:

- **The first box** in the feature control frame is a place for **geometric tolerance symbol**, we learn earlier in this chapter.
- **The second box** is the largest one. It contains three elements:
  - ○ **Symbol of diameter** when you have circular tolerance zone. Remember the first example in the first chapter where the conflict occurred due to the difference in

understanding of tolerance zone as a square zone or circular zone? To avoid any such conflict, it is recommended to put this symbol whenever applicable. In many cases, like flatness, circular tolerance zone does not apply. So this symbol becomes options.

o Next comes the **geometric tolerance value** provided to the feature. Unit of dimension is not mentioned.

o This is the place for any modifier to be considered when assigning a geometrical tolerance.

• Third, fourth and fifth boxes are placeholders for primary, secondary, and tertiary datums, respectively. Datums are non mandatory in many cases (e.g., form and profile tolerances). Therefore, last three boxes are optional.

Based on the above explanations, if we exclude all optional components from feature control fare then the smallest example may look something like given below:

Figure 7.9– Example of the shortest feature control frame.

Important point

Geometric tolerances are provided in addition to dimensional tolerances, as learnt earlier. For the same reason, in practice, complete dimensional and tolerance details would appear as shown below in Figure 7.10:

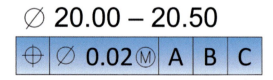

Figure 7.10– Geometric and dimensional tolerances.

7.4 READING FEATURE CONTROL FRAME

You learned how to write in GD&T language. Now let's see how to read it. Let's take the same example from the figure above (7.10).

We will read it out as,

This <mention the feature pointed to> feature has <GD&T symbol name> of <geometric tolerance with diameter > in reference to Primary Datum A, Secondary Datum B, Tertiary Datum C, with Maximum Material Condition.

Altogether, we will speak as follows

This hole feature **has** True Position of 20 Micron Diameter Tolerance in reference to Primary Datum A, Secondary Datum B, Tertiary Datum C, with Maximum Material Condition.

In this chapter, we learnt about:
- Name and symbols of Fourteen geometric tolerances
- Grouping of geometric tolerances in five tolerance types
- Application of fourteen geometric tolerances as independent or in a combination with a datum
- Need of datum for geometric tolerances
- Material condition modifiers (MMC, LMC, RFS)
- Writing in feature control frame
- Reading feature control frame

As mentioned at the beginning of the chapter, deep concepts of fourteen geometric tolerances are pending and being taken in next chapter but, if you are unclear about any concept mentioned above then first clear it before proceeding to next chapter.

8 GD&T TOLERANCES - "FORM"

As promised in the previous chapter, we are starting the detailed study of geometric tolerances from this chapter onwards. We will start with four tolerances as identified in **FORM** tolerance type, as identified in an earlier chapter and again listed below for ready reference.

#	Geometric Tolerance	Symbol	Tolerance Type	Datum required?
1	Straightness	——		Datum is not required because features are measured independently
2	Flatness	▱	Form	
3	Circularity	◯		
4	Cylindricity	⌭		

For each tolerance, we will get into details to cover the following points:
1. Meaning of the geometric tolerance
2. Graphical illustration to clear our understanding
3. Usage of the geometric tolerance and examples
4. Need of datum of the geometric tolerances
5. The relevance of the geometric tolerance with LMC/MMC
6. Writing the geometric tolerance in feature control frame
7. The inclusion of other geometric tolerance(s)

Let's understand the **inclusion** of a geometric tolerance(s) in another geometric tolerance. Let's consider any perfectly flat surface. All lines on the surface will be straight. Therefore, if we provide flatness requirement then we don't need to explicitly mention about straightness requirement. In this case, we say, flatness requirement includes straightness requirement, means flatness in inclusive of straightness.

As we analyze a geometric tolerance, we will note down its details in the format shown in figure 8.1 such that we can build our big picture by the end of this chapter.

Geometric Tolerance	Symbol	Usage	Datum	MMC LMC	Inclusion
Straightness	——	Axis Surface	No	Yes (for axis)	None

Figure 8.1 – Analysis output of straightness geometric tolerance

Are you ready for this exciting journey? Yes! Then let's roll…

8.1 STRAIGHTNESS

Straightness is about a one-dimensional straight line, which is supposed to have the only length. It should not be curved or bent. Refer Figure 8.2, given below.

Straight wire   Curved/bent wire

Figure 8.2 – Illustration of straightness

Let us try to relate straightness with our mechanical industry. You should have seen the hydraulic system in earth moving machine which controls their arms movement, as shown in Figure 8.3, given below.

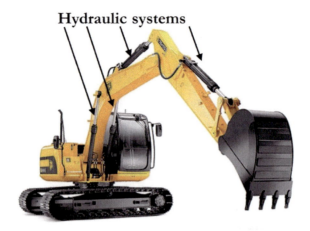

Figure 8.3 – A hydraulic system on earth moving the machine

Figure 8.4 – Hydraulic system close view

Now, look at the design of the hydraulic system as given in figure 8.5.

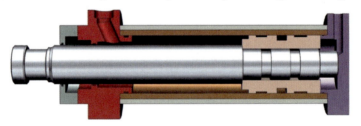

Figure 8.5 – Section view of hydraulic system design

Here your design requirement is to provide a straight rod. It will not only ensure the negligible gap between rod and seal to avoid any oil leakage but also it would avoid any buckling failure.

Now, look at Figure 8.6, given below, for two possible cases for straightness variations.

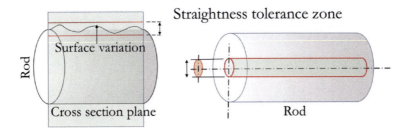

Figure 8.6 – Straightness variation in a rod

There are two profiles on the rod which need to be straight. One is straightness of the surface (shown on the left side) and another is straightness of the axis of the rod (shown on the right side).

- **Surface straightness:** It is a two-dimensional tolerance zone shown between two red lines on a cross section plane. The line profile must remain between these two lines.
- **Axis/line straightness:** It should be a 3D tolerance zone around an ideal axis as shown in right side of the picture as a green cylinder with red boundary.

IMPORTANT NOTE: In ASME Y14.5-2009, and various other sources, this tolerance zone is shown as a 2D zone, similar to the zone shown for surface tolerance. I am aware of this explanation but I do not agree to that. My logic is that deviation cannot be limited to any plane as deviation can occur in any direction. I verified it from the industry and found my understanding to be correct. When the deviation is checked in real life, the deviations are found in all directions, which forms a 3D zone. Therefore, for practical purposes, you should consider the straightness tolerance zone for the axis to be of a 3D shape, however, if you are appearing for any certification exam based on ASME Y14.5-2009 then you should answer in compliance of it, that is you select the 2D option.

§ **Datum for straightness** §
Since straightness is not related to any other feature, so we don't need any datum to provide straightness tolerance.

§ **MMC/LMC for straightness** §
For surface straightness, we don't need material condition but for axis

straightness, we need MMC condition such that for the biggest shaft (or smallest hole) the corresponding fits work as expected.

§ **Writing straightness tolerance** §

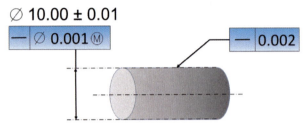

Figure 8.7 – Example of writing straightness tolerance

§ **Inclusion of other tolerances** §
Since it is basic shape so it does not include any other tolerance.

§ **Analysis output of straightness** §
Below is the analysis output of the above learning for straightness:

Geometric Tolerance	Symbol	Usage	Datum	MMC LMC	Inclusion
Straightness	——	Axis Surface	No	Yes (for axis)	None

8.2 FLATNESS

Flatness is about two dimensional flat surfaces like a wall, side of a box, floor, etc., which are called plane in geometry and has two dimensions, length, and breadth. It should not have pits and mounds.

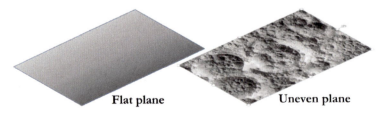

Flat plane   Uneven plane

Figure 8.8 – Illustration of flatness

Let us try to relate flatness to our mechanical industry. Look at the press machine shown in figure 8.9, given below:

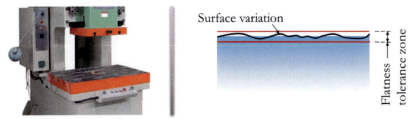

Figure 8.9 – Press machine and flatness illustration

It is important to keep both sides of the press machine as flat as possible such that the output of the product generated after pressing with this machine is of high quality. Look at right side of Figure 8.9 which is an expanded view of the top surface of the bottom pressing plate. The two red lines provide the flatness tolerance zone for the surface.

This particular tolerance is useful in **saving manufacturing cost** as well.

Let's take an example of 20 mm thick tabletop where required flatness precision is 0.10 mm and thickness tolerance is 1 mm. We can control it with dimension tolerance as 20 ± 0.05 mm to meet both the requirements. This design is tough to achieve in the manufacturing process of the entire tabletop.

Alternatively, we can have size tolerance as 20 ± 0.50 mm and flatness tolerance as 0.10. look at figure 8.10, given below.

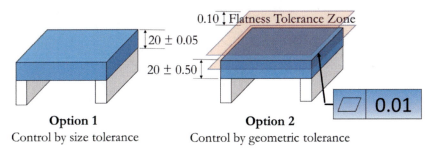

Figure 8.10 – Press machine and flatness illustration

With this design, we can make a tabletop with 0.50 mm tolerance (which is much easier to make as compared to the tabletop with 0.05 mm tolerance) and then we can finish tabletop to achieve 0.10 mm flatness tolerance.

§ Datum for flatness §

Just like straightness, flatness is not related to any other feature, so we don't need any datum to provide flatness tolerance.

§ MMC/LMC for flatness §

The material condition is not so relevant here. It may be needed if we want flatness at a particular plane. This is the reason why the material condition is **optional** for flatness. We will take two cases of flatness, one without material condition and another with MMC.

§ Writing flatness tolerance §

We saw the writing of flatness tolerance without material condition in Figure 8.10. Now we will take flatness at MMC as shown in Figure 8.11.

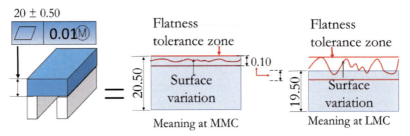

Figure 8.11 – Flatness tolerance with the material condition

§ Inclusion of other tolerances §

Any line on a flat plane will be a straight line. So when you provide flatness tolerance on a surface then you don't need to provide straightness tolerance for any line profile on the same surface. Isn't it included already? Yes. So we can say flatness includes straightness.

§ Analysis output of flatness §

Below is the analysis output of above learning for flatness:

Geometric Tolerance	Symbol	Usage	Datum	MMC LMC	Inclusion
Flatness	▱	Surface	No	Optional	Straightness

Figure 8.12 – Analysis output of flatness

§ <u>Special Case Of Straightness and Flatness</u> §

As a designer, you allowed straightness or flatness or both tolerances by looking at the practicality of the manufacturing process. It resulted in the abrupt rise and drop of the line of plane profiles as shown in Figure 8.13.

Case 1: Gradual surface variation   **Case 2**: Abrupt surface variation

Figure 8.13 – Intended and actual profile variations

To understand it better, let me take help of profile of the roads from the civil industry. Look at Figure 8.14, given below. The first image (8.14a) shows the intention of the engineer to allow some variation in horizontal and vertical directions due to the available terrain of the area.

Figure 8.14a – Intended profile variations of a road

The length of the road was 1 km and engineer allowed total vertical variation of 20 meters, he expected maximum 2 to 3 ups and downs. Similarly, he allowed a maximum horizontal deviation of 50 meters with 2 to 3 turns.

The contractor got the requirement to keep vertical variation within 20 meter and horizontal variation of 50 meter and build roads. It resulted in roads shown in figure 8.14b.

Case 1: Abrupt straightness variation   Case 2: Abrupt flatness variation

Figure 8.14b – Actual profile variations of a road

As you can find in the above picture, if variations intended for longer length is allowed in shorter length then bumps and turns are more, which are undesirable.

The same situation may occur in mechanical design when you want to control straightness and flatness variation over a length or area to avoid sharp bumps or turns.

Consider cases of a vacuum base machine and car shield phone holder mount, which also has a vacuum base to hold on to the mounting surface as shown in figure 8.15, given below:

Figure 8.15 – Vacuum based machine and car phone holder

GD&T provides you with an option to control straightness or flatness over a unit length or area. Suppose you are fine with a total surface variation of 1 mm over 50 mm^2 bases of the device but you don't want more than 0.1 mm variation in 1 mm^2 area to ensure proper functioning of vacuum base. You can write this requirement as shown below in figure 8.16 (A). You can also give area is a circular form with a diameter associated with it, as shown in figure 8.16 (B).

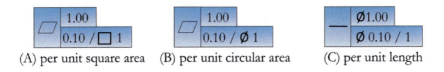

(A) per unit square area   (B) per unit circular area   (C) per unit length

Figure 8.16 – Examples of per unit flatness and straightness

The same concept is applicable to flatness. Refer to Figure 8.16 (C) which allows axial straightness tolerance of 1 mm for a cylinder axis for entire length with the limitation of maximum 0.10 mm variation for 1 mm length.

Per unit tolerance concept is applicable to only straightness and flatness, as per ASME Y14.5-2009. I believe it can be extended to other tolerances as well. Maybe for 'Total runout'? When you learn about other tolerances then you should think if per unit concept can be applied to them or not. This is called progressive thinking!

8.3 CIRCULARITY

Circularity tolerance is relevant to the profile obtained by a cross-section of a cylinder cut by a plane. It provides a two-dimensional tolerance zone between two circles as shown below in Figure 8.17.

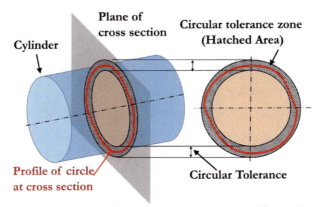

Figure 8.17 – Circularity geometric tolerance illustration

Circularity is also called roundness. It is the most common geometric tolerance applicable to rotating parts like shaft. Circularity is particularly important for a cross-section area where you have sealing between shaft and hole to avoid any oil leakage or to minimize wear and tear. You can imagine other examples of the shaft connecting engine and generator, etc.

Circularity is also useful in **saving manufacturing cost,** just as you saw in the case of flatness to save manufacturing cost of the table by shifting dimension tolerance of table to flatness tolerance of tabletop. Similarly, for circularity also, we may increase diameter tolerance and add circularity tolerance for economical manufacturing.

§ **Datum for circularity** §
Since circularity is not related to any other feature, just like straightness and flatness, we don't need any datum to for circularity tolerance.

§ **MMC/LMC for circularity** §
The material condition is not required for circularity. It is always taken in relation to the top surface of the shaft.

§ **Writing circularity tolerance** §
Writing circularity is very simple as it does not have material condition or datum associated with it. It is written as shown in Figure 8.18.

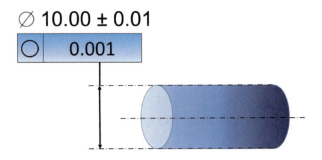

Figure 8.18 – Writing circularity tolerance

§ **Inclusion of other tolerances** §
Since the circle is a basic shape, therefore circularity does not include any other geometric tolerance.

§ Analysis output of circularity §

Below is an analysis output of above learning for circularity:

Geometric Tolerance	Symbol	Usage	Datum	MMC LMC	Inclusion
Circularity	◯	2D cross section	No	No	None

Figure 8.19 – Analysis output of circularity

8.4 CYLINDRICITY

Cylindricity tolerance is an extension of circularity tolerance throughout the length of a cylinder. It is a combination of circularity and surface straightness. It provided a tolerance zone made of space between two cylinders as shown below in Figure 8.20.

It is a common geometric tolerance applicable to any part which needs to be both round and straight, for example sliding shafts, pins and any critical cylindrical element bush and housing, etc.

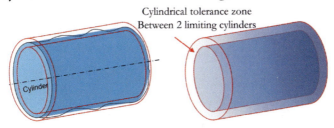

Figure 8.20 – Cylindricity geometric tolerance illustration

Refer to Figure 8.21, given below for an example of bush and housing.

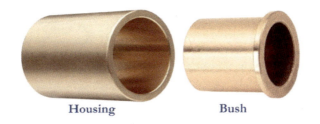

Figure 8.21 – Cylindricity geometric tolerance example

Here both bush and housing need to be perfectly cylindrical to fulfil the requirement. Manufacturing up to such precision becomes difficult during mass production but it becomes feasible and cost-effective to produce with more tolerance and then machining to achieve geometric cylindrical tolerance. Therefore, cylindricity also **saves manufacturing cost.**

§ **Datum for cylindricity** §
Since cylindricity is not related to any other feature, just like straightness and flatness, we don't need any datum to it.

§ **MMC/LMC for cylindricity** §
The material condition is not required for cylindricity. It is always taken in relation to the top surface of the shaft.

§ **Writing cylindricity tolerance** §
Writing cylindricity is is very simple as it does not have any associated material condition or datum. It is written as shown in Figure 8.22.

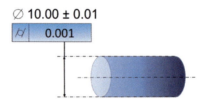

Figure 8.22 – Writing circularity tolerance

§ **Inclusion of other tolerances** §
Since the cylinder is an extension of a circle along a straight line, defining cylindricity would include circularity and straightness.

§ **Analysis output of circularity** §
Below is an analysis output of above learning for cylindricity:

Geometric Tolerance	Symbol	Usage	Datum	MMC LMC	Inclusion
Cylindricity	/⌀/	Straightness Roundness	No	No	Straightness Circularity

Figure 8.23 – Analysis output of circularity

Chapter recap

In this chapter, we learnt about four geometric tolerances as part of FORM tolerance type. These tolerances are:

- Straightness, flatness, circularity, and cylindricity.
- We did a detailed analysis, looked at different examples, and consolidated our learning in the below form:

Geometric Tolerance	Symbol	Usage	Datum	MMC LMC	Inclusion
Straightness	——	Axis Surface	No	Yes (for axis)	None
Flatness	▱	Surface	No	Optional	Straightness
Circularity	○	2D cross section	No	No	None
Cylindricity	⌭	Straightness Roundness	No	No	Straightness Circularity

- We also learnt about the special case for straightness and flatness to define per unit tolerance.
- We learnt about writing above tolerances in GD&T format

If any point is unclear, then clarify it before we proceed to PROFILE type tolerances.

9 GD&T TOLERANCES – "PROFILE"

In the previous chapter, we learnt about tolerances related to basic forms of line, plane, circle, and cylinder. In this chapter, we will continue our journey to cover two more tolerances grouped under PROFILE tolerance class as listed in chapter 7 and again listed below for ready reference.

#	Geometric Tolerance	Symbol	Tolerance Type	Datum required?
5	Line Profile	⌒	Profile	Datum may or may not be needed (optional)
6	Surface Profile	⌓		

If you look at the exterior surface of cars, aeroplanes, motorbikes, boats, etc. then you would find lots many other shapes or combination of shapes. These curves and shapes serve the purpose of least resistance by reducing aerodynamic forces and also improving the aesthetic look of the exterior surface. Let's have some leisure time watching these wonderful shapes and profile in Figure 9.1.

Figure 9.1 – Examples of surface profiles in real life

These surface designs are good to look at and at the same time extremely difficult to design. The biggest challenge is to communicate the profile details to the manufacturer in such a way that the final output matches the intended design. This is done with the help of two profile tolerances in GD&T which we are going to learn in this chapter.

9.1 PROFILE OF A LINE

Generally, curved surfaces are complex in nature and need careful profiling. Let's take an example of an aeroplane wing and study line profile created at the top surface, at a cross-section, as shown in Figure 9.2:

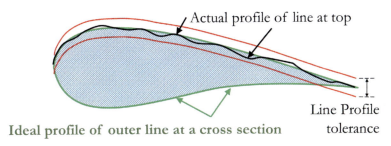

Figure 9.2 – Line profiling of aeroplane wing at a cross-section

Here you may find a green curved line around the hatched area which shows the ideal profile the designer would like to get manufactured. Now concentrate on the upper surface and find black line representing actual profile of line at the top of wings. Two red lines are given by the designer to keep the actual line profile between these two red lines.

Let us take another real-life example, given below in figure 9.3. It is a cross-section of a plastic part in a car.

Figure 9.3 – Line profiling of plastic part of a cross-section

The green line shows the desired profile of the line at the cross-section. Set of 2 red lines are to define the tolerance for line profile.

After looking at above two examples, we can say, a line profile is a 2D profile of a curved line over a curved surface, taken at a cross-section.

Let's move forward to understand other details of the tolerance.

§ Datum for a profile of a line §

You may like to control the profile of a line in conjunction with other features or datum. It is particularly important when you want continuity of flow of contour between two surfaces. For example, a line profile may flow from bonnet of the car to the front door to back door to all the way to the backside of the car. Look at Figure 9.4, given below where one curved profile (purple) is shown on a block (green). On the purple curved profile, one black line is shown which is a profile of a line to which we intend to assign tolerance in reference to datum A and B.

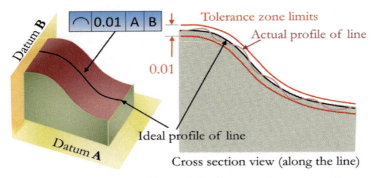

Figure 9.4 – Line profiling of plastic part of a cross-section

Tolerance gave to profile if the line is 0.01. Refer to the right side of the picture, which is cross-sectional view along the line. The black line is ideal profile, the purple line is an actual profile and red lines show the tolerance zone within which actual profile should be limited to exist. There may be a situation when the entire surface is profiled independently. In such cases, datums are not required.

§   **MMC/LMC for a profile of a line**   §
The material condition is not required for a profile of a line as it is always taken at the top surface of a curved surface profile.

§   **Writing profile of a line**   §
Writing a profile of a line is very simple as shown in Figure 9.4, given above. It can only go simpler when datums are not required.

§   **Inclusion of other tolerances**   §
Profile of a line is similar to straightness. The only difference is that it is curved in a 2D plane. Therefore, it does not include any other tolerance.

§   **Analysis output of profile of a line**   §
Below is analysis output of above learning for a profile of a line:

Geometric Tolerance	Symbol	Usage	Datum	MMC LMC	Inclusion
Profile of a line	⌒	2D line profile	Optional	No	None

Figure 9.5 – Analysis output of profile of a line

9.2 PROFILE OF A SURFACE

It is a 3D profile of a curved surface. Refer to figure 9.6, given below, in which a car hood picture is shown on left figure.

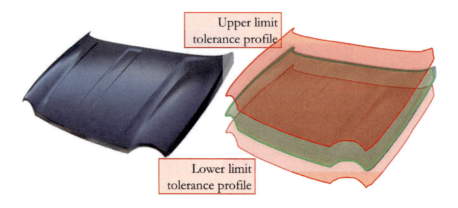

Figure 9.6–Example and illustration of the profile of a surface

Illustration on the right is surface profiling of same car hood. The ideal surface profile is shown in green color and tolerance profiles are shown in red color, showing upper limit and lower limit tolerance profile.

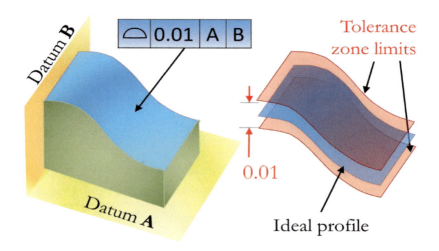

Figure 9.7–Illustration of the profile of a surface and notations

Look at Figure 9.7, given above, in which blue top of the curved profile is given a tolerance of 0.01 mm. The tolerance zone is shown by two red curved surfaces on the right side of the picture. The actual profile should remain between these two red planes.

§ Datum for a profile of a surface §

Just like the profile of a line, you may like to control the profile of a surface in conjunction with other features or datum, however, you may have surfaces which are independently controlled. Therefore, again datum becomes optional.

§ MMC/LMC for a profile of a surface §

The material condition is not required for a profile of a surface as it is always taken at the top surface of a curved surface profile.

§ Writing profile of a surface §

Writing a profile of a surface is very simple as shown in figure 9.7, given above. It can only go simpler when datums are not required.

§ Inclusion of other tolerances §

Profile of a surface is similar to flatness. The only difference is that it is curved in a 3D plane. As flatness includes straightness, the profile of a surface will include the profile of a line Look at Figure 9.8 which shown multiple paths, as black lines, on the curved surface of the profile. Any profile checking equipment will measure the accuracy of profiling by travelling through these paths on actual surface.

Figure 9.8–Paths to check the profiling on a profile of a surface

These paths are along a curved line obtained as a cross-section of the curved profile surface with a plane. Obviously, it shows the profile a line on the same surface. Therefore, when we provide a profile of a surface, it is inclusive of the profile of a line.

§ Analysis output of profile of a line §

Below is analysis output of above learning for a profile of a surface:

Geometric Tolerance	Symbol	Usage	Datum	MMC LMC	Inclusion
Profile of a surface		3D curved surface	Optional	No	Profile of a line

Figure 9.9 – Analysis output of the profile of a surface

❓ Question time !

If the profile is flat, then the profile of the surface will include flatness. Right? Then we did not include flatness in inclusion. Did we make a mistake and shall we revise above understanding?

The answer is "No".

Although the profile of a surface can be used to control the flatness, we don't use the profile of a surface for such purposes. Profile of a surface can cover a wide range of profile and almost all profiles can be controlled by it. However, we use it when nothing else can be used.

9.3 SPECIAL CASE OF PROFILE – NON-UNIFORM ZONE

Till now we always found a uniform zone of tolerances created by any geometric tolerance studied so far. Recall straightness in which the zone was created between two parallel lines or cylindrical zone at axis, for flatness, the zone was formed between two parallel planes, for circularity, the zone was created between two concentric circles, which is again uniform, and for cylindricity, it was between two coaxial cylinders, which is again uniform. In fact, you will find it to be the same case for all remaining profiles we are going to learn in this book.

However, our "**Profile**" tolerance has got special consideration in the 2009 edition of ASME standard. It can have varying or non-uniform tolerance zone. Look at one example given below in Figure 9.10:

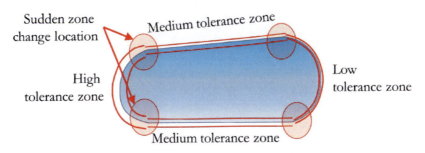

Figure 9.10 – Illustration of varying tolerance zone of a profile

The feature has a requirement of less tolerance on the right side, medium tolerance in the central part and can allow bigger tolerance on the left side, as shown by red lines. You can see sudden tolerance zone area change at four places highlighted by red circles. This kind of disruption in tolerance zone causes manufacturing issue and raise confusion between design and manufacturing teams.

Now, look at another example in Figure 9.11, given below, in which connection issue is not there but designer wants to provide an increased tolerance zone in the centre area of the feature to increase manufacturing convenience and reduce manufacturing cost.

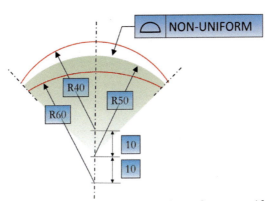

Figure 9.11 – Example and representation of a non-uniform zone

Here writing in GD&T is simple, just put symbol and write NON-UNIFORM next to it. But it is insufficient for the detailed purpose. You have to put all geometric detail in drawing or put notes in the drawing.

Chapter  recap

In this chapter, we learnt about two geometric tolerances as part of PROFILE tolerance type namely profile of a line and surface. We did a detailed analysis, looked at different examples, and consolidated our learning in the below form:

Geometric Tolerance	Symbol	Usage	Datum	MMC LMC	Inclusion
Profile of a line	⌒	2D line profile	Optional	No	None
Profile of a surface	⌓	3D curved surface	Optional	No	Profile of a line

We also learnt about the special case for a profile to be non-uniform. If any point is unclear then clarify it before we proceed to ORIENTATION type tolerances.

10 GD&T TOLERANCES – "ORIENTATION"

Meaning of word orientation is "action of placing something relative to other specified position". It means Orientation is always relative. In this chapter, we will continue our journey to cover three more tolerances grouped under ORIENTATION tolerance class as listed in chapter 7 and again listed below for ready reference.

#	Geometric Tolerance	Symbol	Tolerance Type	Datum required?
7	Angularity	∠		
8	Perpendicularity	⊥	Orientation	Datum is mandatory to mention
9	Parallelism	//		

Before we dive into these tolerances, I would like to establish two concepts which will make this learning a baby's game.

1ST CONCEPT:

Angular tolerance is different from angularity tolerance. Look at figure 10.1, given below: in which angular tolerance denotes a variation in the angle of inclination between two lines or planes.

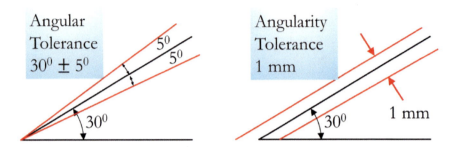

Figure 10.1 – Difference between angular and angularity tolerance

On the other hand, angularity tolerance creates a uniform thickness zone around ideal orientation in which line or plane may remain.

2ND CONCEPT

Perpendicularity and parallelism are an extension of angularity with the angle of inclination being 90^0 and $0^0/180^0$ as shown below in figure 10.2.

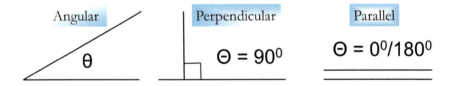

Figure 10.2 – Angular in form of perpendicular and parallel

After understanding these two concepts, you will smoothly sail through the orientation tolerances. Let's move forward…

10.1 ANGULARITY

Now when you look at figure 10.3, given below, for angularity tolerance zone, then you can understand each bit of it. Isn't it?

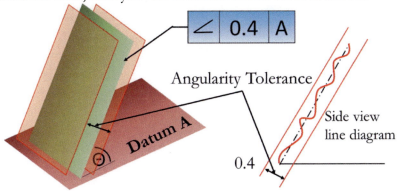

Figure 10.3 – Illustration of angularity tolerance

Clearly, the red transparent planes are forming the tolerance zone for the green ideal plane. Picture of the right side is line diagram for same.

It's your turn to imagine how would angularity for a line work? What would be the shape of such tolerance zones?
Did you visualize a cylindrical tolerance zone, which is the case for straightness as well?

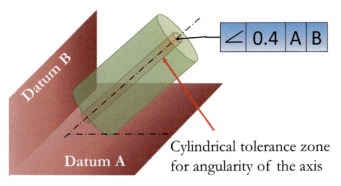

Figure 10.4 – Illustration of angularity tolerance for a line

Look at Figure 10.4, given above, illustrating the same case. I am sure it was easy for you to understand it *(else revisit section 8.1)*

 Question time !

The above simple question was a trap for you, the real question is coming now. Tell me the importance of datum B. When angularity is given with respect to datum A then what difference datum B is making here?

Hint: Answer is related to the datum reference frame and degree of freedom. Refer to section 6.1 and try to answer this question.

§ **Datum for angularity** §
As explained at the beginning of this chapter, all orientation tolerances are relative to one or more datum. So datum is applicable.

§ **MMC/LMC for angularity** §
You can apply a material condition to angularity but it is conventionally not used to keep it simple. In the next section, you will find it to be useful for perpendicularity tolerance.

§ **Writing angularity** §
Writing angularity is very simple as shown in figures 10.3 and 10.4.

§ **Inclusion of other tolerances** §
As it controls axis and plane, straightness and flatness are included.

§ **Analysis output of angularity** §
Below is an analysis output of above learning for angularity:

Geometric Tolerance	Symbol	Usage	Datum	MMC LMC	Inclusion
Angularity	∠	Axis Surface	Yes	Yes Less used	Straightness Flatness

Figure 10.5 – Analysis output of angularity

78

10.2 PERPENDICULARITY

Perpendicularity is a special case of angularity where able between plane and datum is 90^0. All other explanation remains the same except the impact of a material condition which we will see later. First, let's look into an illustration of perpendicularity in figure 10.6, given below:

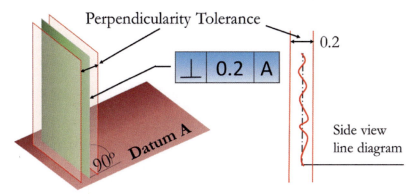

Figure 10.6 – Illustration of perpendicularity

Here we have mentioned only one datum. Can you guess when two datums may be used? Yes, obviously when you need two plane references. Look at Figure 10.7, given below, to understand it. on the left side of the figure, we used only one datum. Perpendicularity tolerance controls it with respect to datum A but allows the plane to rotate, as shown, in such a manner that it remains perpendicular to datum A. It does not meet our requirements. To restrict this rotational degree of freedom, we apply another datum B, as shown on right side of Figure 10.7.

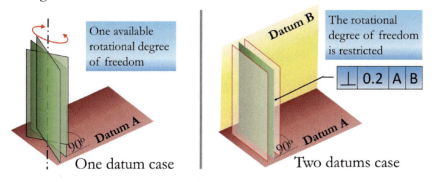

Figure 10.7 – Illustration of perpendicularity with 2 datums

As a result of datum B, the remaining degree of freedom is restricted as we wanted. In such cases, we use two datums.

 Question time !

Can we extend the application of datum to three datums to control perpendicularity? If yes, then what be the case when the third datum will be required and how will we apply the third datum?

I think I should answer this rather than leaving it to you.

Answer: You cannot have a plane which is perpendicular to three other mutually perpendicular planes (as three planes of datum reference frame). Geometrically it is not possible. Try to place a plane in the datum reference frame but you would not be able to do it. Therefore perpendicularity cannot have three datums.

 Question time !

How do we decide if we need to apply no datum, one datum or two datums for perpendicularity?

Answer: It is driven by the available degree of freedom. If no degree of freedom is available, it may be constrained by other features, then we don't need to apply any datum. If one degree of freedom is available, then apply one datum and if two degrees of freedom are available then apply two datums.

Do you remember one unanswered question asked in angularity? The answer to that question is also to control the degree of freedom by referencing to datum B.

§ MMC/LMC for perpendicularity §

This next concept is very important to understand and needs proper attention. It will also create a foundation for another important concept of **bonus tolerance**. So, proceed once you are ready.
Consider the design of a pin and hole using the Hole Basis System. Considering it to be clearance fit, basic size 16 mm, 0.1% fundamental

deviations on hole and shaft (=0.016 mm). We select IT7 grade fit which is the moderate grade for a clearance fit with a tolerance of 0.018 mm (later we will learn in detail about IT grade system to calculate dimensional tolerances). Now let us calculate MMC sizes for hole and shaft.

- MMC diameter of hole = Basic size + fundamental deviation + dimensional tolerance = 16.000+0.016+0.018 = 16.034 mm.
- MMC diameter of shaft = basic size – fundamental deviation
  = 16.000 - 0.016 = 15.984

Look at Figure 10.8 to visualize an MMC condition interactions. The left image does not consider perpendicularity tolerance and the right image illustrates the situation with maximum perpendicularity tolerance.

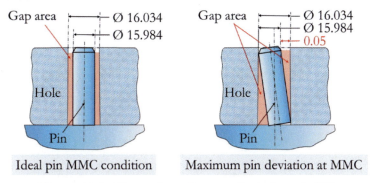

| Ideal pin MMC condition | Maximum pin deviation at MMC |

Figure 10.8 – Illustration of perpendicularity at MMC

Closely look at picture details Red shaded areas are a gap between pin and hole. The total available gap to pin at MMC =

MMC of hole – MMC of shaft = 16.034 – 19.984 = 0.050 mm.

This is maximum allowed tolerance for the pin to be utilized for any additional tolerance, which may be additional dimensional tolerance or any geometric tolerance. In this case, we are using it for perpendicularity. So we can say, maximum perpendicularity tolerance available is 0.050 mm at MMC.

⇨ This is the way we derive tolerance

Now we will at another important concept of impact on tolerance at LMC. Look at Figure 10.9, given below to take the above case to see the derived tolerance at LMC of a pin.

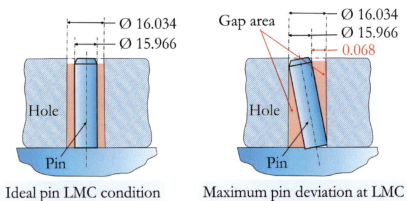

Ø 16.034
Ø 15.966

Gap area

Ø 16.034
Ø 15.966
0.068

Hole

Hole

Pin

Pin

Ideal pin LMC condition   Maximum pin deviation at LMC

Figure 10.9 – Illustration of perpendicularity at LMC

Pin at LMC reduced to MMC – dimensional tolerance based on IT grade which is equal to $15.984 - 0.018 = 15.966$, as shown above. In the left image, you may find the maximum possible gap between pin and hole at LMC = hole diameter at MMC (hole basis system) – pin diameter at LMC = $16.034 - 15.966 = 0.068$ mm.

Now, look at the right image. The pin can have a maximum deviation to the extent of the maximum available gap which is 0.068 mm.

Extra tolerance you got between MMC and LMC = $0.068 = 0.050 - 0.018$. It is same as the difference between MMC and LMC of the pin. It is not a coincidence. It is mathematics. It is engineering. You get an extra tolerance of the same amount as much as you move away from MMC up to LMC. Look at Figure 10.10 to validate this concept.

Hole Diameter (Fixed at MMC)	Pin Diameter	Allowed Tolerance (hole dia.–pin dia.)	Deviation from MMC	Extra tolerance
16.034	15.984	0.050	0.000	0.000
16.034	15.983	0.051	0.001	0.001
16.034	15.982	0.052	0.002	0.002
.	.	.	.	.
.	.	.	.	.
16.034	15.968	0.066	0.016	0.016
16.034	15.967	0.067	0.017	0.017
16.034	15.966	0.068	0.018	0.018

Figure 10.10 – Extra tolerance due to deviation from MMC

⇨ This extra tolerance is called bonus tolerance.

The concept of bonus tolerance is very important for the manufacturer as they use this bonus tolerance to manufacture more output with less rejection and less cost due to increase tolerance. As a designer, you should understand it when you provide tolerance at MMC.

Coming back to our original point, if the material condition is applicable to perpendicularity, now you can see it is definitely applicable.

§ **Writing perpendicularity** §
Figure 10.11 shows different cases for writing tolerance.

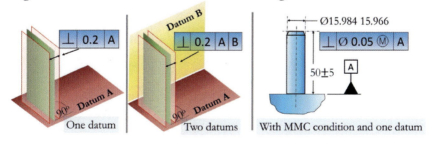

Figure 10.11 – Writing perpendicularity tolerances

§ **Inclusion of other tolerances** §
As it controls axis and plane, straightness and flatness are included.

§ **Analysis output of perpendicularity** §

It is almost the same as angularity. The only difference is MMC becoming important in case of perpendicularity.

Geometric Tolerance	Symbol	Usage	Datum	MMC LMC	Inclusion
Perpendicularity	⊥	Axis Surface	Yes	Yes	Straightness Flatness

Figure 10.12 – Analysis output of perpendicularity

10.3 PARALLELISM

Parallelism will be simpler to understand. It is mostly used to mention parallelism between two parallel planes. In such cases, only one datum is used and the material condition is not required. Look at Figure 10.13, given below, to understand its application and method of writing.

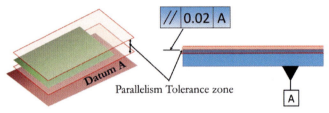

Figure 10.13 – Illustration of parallelism between planes

Important note: Flatness is not the same as parallelism. Parallelism uses a datum to control a surface while flatness does not. If a plane is flat, it need not be parallel to any datum. If we need to control the flatness, then tolerance will modify as given below in Figure 10.14.

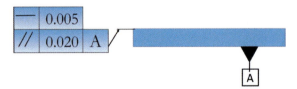

Figure 10.14 – Illustration of parallelism with flatness

Mainly parallelism is used for geometric conditions, as shown above, that is, to define parallelism between two planes while making reference plane at datum. Now we will discuss two more cases which are slightly less common which are used for parallelism of an axis. Obviously, the tolerance zone of an axis will be cylindrical as we will see.

1. **Parallelism with two datums**. Look at Figure 10.15, given below, in which we are giving parallelism tolerance to the axis of a hole in the centre of a cubical block. Here the axis needs to be parallel to two datums, A and B.

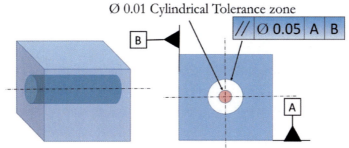

Figure 10.15 – Example of parallelism with 2 datums

2. Parallelism with a material condition: Taking a case of connecting rod in which bigger hole is taken as a datum (with RFS) and smaller hole is taken at MMC. Look at Figure 10.16, given below, to understand the concept.

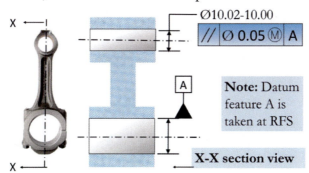

Figure 10.16 – Example of parallelism with the material condition

Here, the lower hole is taken as reference datum at RFS and the upper hole is given parallelism tolerance at MMC. Here again, the concept of bonus tolerance will be applicable.

§ Inclusion of other tolerances §

As it controls axis and plane, straightness and flatness are included.

§ Analysis output of parallelism §

It is almost the same as angularity. The only difference is MMC may become important in case of parallelism.

Geometric Tolerance	Symbol	Usage	Datum	MMC LMC	Inclusion
Parallelism	//	Axis Surface	Yes	Yes	Straightness Flatness

Figure 10.17 – Analysis output of parallelism

Chapter  recap

In this chapter, we learnt about three geometric tolerances as part of ORIENTATION tolerance type namely angularity, perpendicularity, and parallelism.

- We understood the difference between angular tolerance and angularity tolerance.
- We learnt two important concepts of **bonus tolerance**
- We saw use cases of two datums and material conditions.
- We did a detailed analysis, looked at different examples, and consolidated our learning in the below form:

Geometric Tolerance	Symbol	Usage	Datum	MMC LMC	Inclusion
Angularity	∠	Axis Surface	Yes	Yes Less used	Straightness Flatness
Perpendicularity	⊥	Axis Surface	Yes	Yes	Straightness Flatness
Parallelism	//	Axis Surface	Yes	Yes	Straightness Flatness

If any point is unclear, then clarify it before we proceed to LOCATION type tolerances.

11 GD&T TOLERANCES – "<u>LOCATION</u>"

In this chapter, we will continue our journey to cover three more tolerances grouped under LOCATION tolerance class as listed in chapter 7 and again listed below for ready reference.

#	Geometric Tolerance	Symbol	Tolerance Type	Datum required?
10	True Position	⊕	Location	Datum is mandatory
11	Concentricity	◎		
12	Symmetry	≡		

This location tolerance type is used to control

- **The position** of a point (e.g. centre of the circle), axis or central plane
- **Concentricity** between the axis of feature and datum axis or point
- **The symmetry** of two features across a datum plane.

Let's start the journey to control locations of features...

11.1 (TRUE) POSITION

*Accurate terminology, as per ASME standard, is "**Position**", however, the **True position** is also widely. used. Therefore, we can use either of them.*

Remember the first GD&T case we discussed in the first chapter? Let's start with a similar case as shown in Figure 11.1, given below.

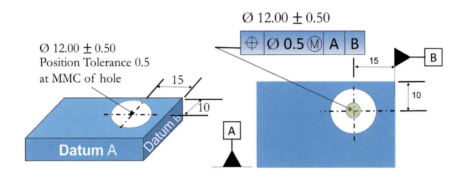

Figure 11.1 – Simple position tolerance example

Look at the left image. Here designer is suggesting a hole of diameter 12 mm, with 0.50 mm dimensional tolerance, with position or the centre at 10 mm from datum A and 15 mm from datum B, with 0.50 mm positional tolerance when the hole is at MMC (smallest hole). The image on the right shows the position tolerance zone in green colour (at the centre of the hole). It means the centre of the hole may remain in any place in the identified area.

Here one point is important to note. **The material condition is MANDATORY for position tolerance**. It can be MMC, LMC or RFS. It can be applied to the feature itself or other datum features. We will look at such cases. Before that, I would like you to understand the impact of MMC material condition applied at feature (hole) itself.

Refer to Figure 11.2, given below. Consider the mating part of the hole is a shaft of diameter 11.50 mm at its MMC, as shown by grey shaded circle. By giving 0.50 mm position tolerance, the designer is ensuring the shaft to fit properly. When a manufacturer produces makes the hole, the hole should be able to accommodate the shaft. This is required to meet the functional need.

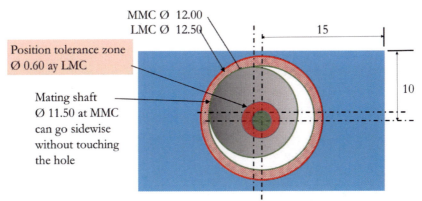

Simplified GD&T

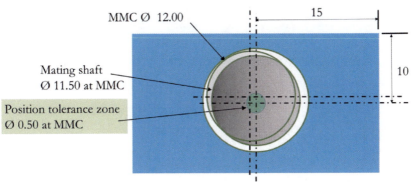

Figure 11.2 – Illustration of fit at MMC

Now refer to Figure 11.3 in which another extreme condition of the hole is illustrated, that is when the hole is made at LMC (biggest hole in red colour). Here manufacturer will get a bigger tolerance zone to ensure proper fit. Because the shaft gets more space to move sideward up to the extent of touching the hole, as needed.

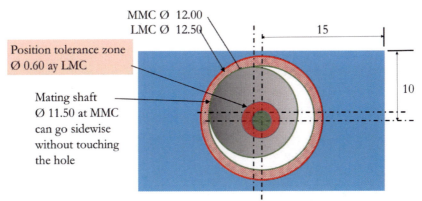

Figure 11.3 – Illustration of fit while departing from MMC to LMC

This extra tolerance is the same as BONUS TOLERANCE as we learnt in the previous chapter when we learnt about perpendicularity for a pin (shaft). Therefore, now the concept of BONUS tolerance should have been clear to you for cases when tolerance is provided at MMC. To summarize, we can say, bonus tolerance is extra tolerance made available to the manufacturer when moving from MMC to LMC The value of extra tolerance will be same as the value of departure from MMC.

If the bonus tolerance concept at MMC is not yet clear, then revisit section 10.2 and again go through the above example to clarify it before you proceed to next topic

 ## Question time !

If position tolerance is given at LMC then will you get a bonus tolerance?

Answer and hint: Bonus tolerance concept is applicable at LMC as well. Consider a case of hole and shaft. Suppose you need to ensure some minimum gap between hole and shaft and you have provided position tolerance at LMC then as you move away from LMC towards MMC, you get extra bonus, of same value as you moved away from LMC to MMC as you need to shift the position of hole away from ideal position to ensure minimum gap. Try to write your own figure and see if you can do it.

§ **Material conditions at datum features** §

As mentioned earlier, the datum is mandatory for position tolerance. It can be applied to the feature or datum features as well. Look at Figure 11.4, given below, in which datums are provided at the feature as well as two datum features A B and C.

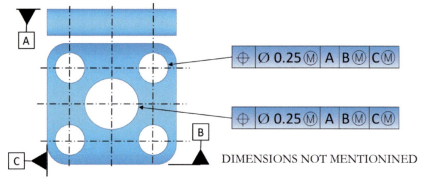

Figure 11.4 – Material condition at the feature and datum features

Here the design is to make one bigger hole at centre and four small holes at corners of a rounded rectangular plate. Look at Figure 11.5, given below, illustrating the situation at LMC.

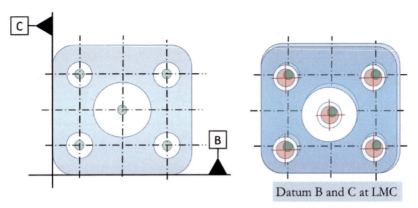

Figure 11.5–Effect of MMC at datum features

When datum features B and C departs from MMC to LMC, position tolerances of holes are increased in order to remain functional with mating part(s). the increased tolerance zones are shown by red circles.

§ **Position tolerance for axis** §

Let's again go back to our first example to see how position tolerance can be used for the axis. Look at Figure 11.6, given below.

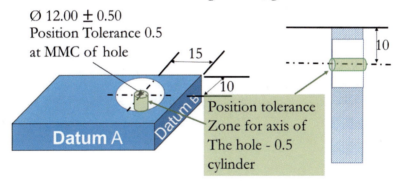

Figure 11.6–Position tolerance for the axis of a hole

Here you can see a green cylindrical zone to control the position of the axis of the hole. It is a three-dimensional control and widely used when the depth of the hole increases such that proper fit can be ensured.

§ **MMC/LMC for position tolerance** §

We learnt that material condition is mandatory for the position. It is also true because position tolerance is always applied to a feature of size.

§ **Writing position tolerance** §

In multiple examples, we learnt multiple cases to write position tolerance. An important role is played by a material condition which can apply to the feature as well as datum features.

§ **Inclusion of other tolerances** §

Since it is basic position tolerance so it does not include any other geometric tolerance.

§ **Analysis output of position tolerance** §

Below is the analysis output for position tolerance.

Geometric Tolerance	Symbol	Usage	Datum	MMC LMC	Inclusion
Position	⊕	Point Axis	Yes	Yes	None

Figure 11.7 – Analysis output of position tolerance

11.2 CONCENTRICITY

Concentricity is a 3D geometric tolerance zone of circular features (like a cylinder, cone, etc.) to be coaxial with reference (Datum) axis, as shown below in Figure 11.8:

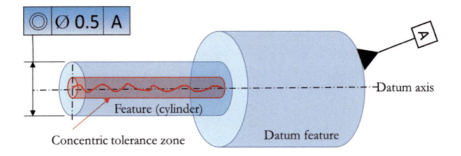

Figure 11.8 – Illustration of concentricity of a cylinder

Here concentricity of the axis of the feature (smaller cylinder) with the axis of the datum feature (bigger cylinder) is controlled by concentricity tolerance, which is 0.5 mm as shown in the figure. It creates a cylindrical tolerance zone as shown by transparent red cylindrical zone around the ideal axis of the feature.

Axes of datum feature and datum are measured by the distribution of the center of mass of these features. It is extremely difficult to measure and demonstrate. Due to this difficulty, concentricity tolerance is rarely used in industry. Instead of it, we can use runout or total runout tolerances. These will be taught in next chapter.

§ **MMC/LMC for concentricity** §
Since it is based on the mass distribution of feature and datum feature, the material condition is not used.

§ **Writing concentricity tolerance** §
Since it is not widespread, so the example we saw above is sufficient.

§ **Inclusion of other tolerances** §
It does not include any other geometric tolerance.

§ **Analysis output of concentricity tolerance** §
Below is the analysis output for concentricity tolerance.

Geometric Tolerance	Symbol	Usage	Datum	MMC LMC	Inclusion
Concentricity	◎	Shafts Fits	Yes	No	None

Figure 11.9 – Analysis output of concentricity tolerance

11.3 SYMMETRY

Refer to Figure 11.10 to understand the meaning of symmetry. In the left side of the figure, you can see a symmetrical object with a plane of symmetry in green colour. When you look at the image on the right side then you would find the object to be asymmetrical across the green plane. There may exist another vertical plane across which object may be symmetrical but we are interested only in our own green plane. It may be our functional requirement.

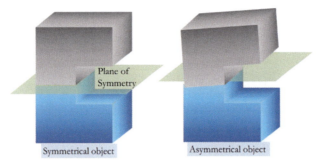

Figure 11.10 – an Illustrative example of symmetry

How is this plane of symmetry identified? Using dimensions? No. GD&T standard expects you to find the centre of mass on both sides of the plane of symmetry and check where it falls. If all such measurement falls on the (green) plane of symmetry then you can say it is perfectly symmetrical. Such quality is hard to achieve so we provide symmetry tolerance to allow some deviations.

Look at Figure 11.11, in which two red planes are representing the upper and lower limits of tolerances which creates a tolerance zone between which centre of mass of all opposite mass should fall.

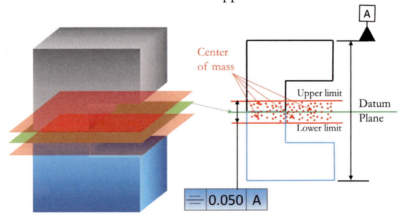

Figure 11.11 –Symmetry tolerance zone and calculation method

Figure on right-hand side in the above figure is a schematic diagram of the same setup. Green plane of symmetry is represented by a green line and red tolerance zone planes are represented by red lines. These centre of masses are represented by red dot between red lines. If all dots are falling within tolerance zone, then the object would be accepted as a symmetrical object.

As you must have got the question, how to measure the center of opposite masses and represent it? Actually, it is not only difficult to measure it but the process becomes expensive too. It leaves this tolerance more theoretical and less practical. So it is seldom used.

So what is a workaround for symmetry?

When you look closely then you would realize the symmetry has the capability of controlling position, flatness, and also parallelism. If you could control symmetry then you would not need to control position, flatness, and parallelism for features already covered in symmetry, Isn't it? We will make use of the same property of symmetry but in reverse order. Means, we will control symmetry by controlling position, flatness, and parallelism of different features. Liked the idea? This is exactly what is being practised in industry. Instead of using symmetry tolerance and getting involved in expensive and complicated measurement techniques, industry uses position, flatness, and parallelism to achieve the same result. This is the workaround.

§ **MMC/LMC for Symmetry** §

Since it is based on the mass distribution of feature and datum feature, so the material condition is not used.

§ **Writing symmetry tolerance** §

Since it is not widespread, so the example we saw above is sufficient.

§ **Inclusion of other tolerances** §

It is a very comprehensive 3D geometric tolerance across a central datum plane across which all features need to be symmetrical. It includes flatness, parallelism, and position.

§ **Analysis output of symmetry tolerance** §

Below is the analysis output for symmetry tolerance.

Geometric Tolerance	Symbol	Usage	Datum	MMC LMC	Inclusion
Symmetry	═══	3Ds symmetry	Yes	No	Flatness Parallelism Position

Figure 11.12 – Analysis output of symmetry tolerance

Chapter recap

In this chapter, we learnt about three geometric tolerances as part of LOCATION tolerance type namely position, concentricity, and symmetry. Below are the main points covered:

- Position tolerance is provided for the feature of size.
- Position tolerance need material conditions.
- The material condition can be provided with the feature of size or datum features or both.
- Position tolerance zone will be a circular zone for position tolerance of a point. It will be a cylindrical zone for axis position.
- Concentricity tolerance is coaxiliaty of the axis of the feature with the axis of the feature with feature axis. It is measured on the basis of the center of mass of feature and datum feature, if applicable.
- Symmetry is most comprehensive 3D tolerancing which works on the concept of center of mass across datum plane.
- Due to complexity to calculate the center of mass, both concentricity and symmetry are rarely used in industry.
- We did a detailed analysis, looked at different examples, and consolidated our learning in below form:

Geometric Tolerance	Symbol	Usage	Datum	MMC LMC	Inclusion
Position	⊕	Point Axis	Yes	Yes	None
Concentricity	◎	Shafts Fits	Yes	No	None
Symmetry	=	3Ds symmetry	Yes	No	Flatness Parallelism Position

If any point is unclear, then clarify it before we proceed to RUNOUT type tolerances.

12 GD&T TOLERANCES – "RUNOUT"

In this chapter, we will continue our journey to cover two remaining tolerances grouped under RUNOUT tolerance class as listed in chapter 7 and again listed below for ready reference.

#	Geometric Tolerance	Symbol	Tolerance Type	Datum required?
13	Circular Runout	↗	Runout	Datum is mandatory
14	Total Runout	↗↗		

RUNOUT is not related to cricket! It is related to something even more popular in the mechanical industry. It is related to something without which mechanical industry cannot exist. It is related to something which is involved in every person's life. **It is related to rotating parts**. Have a look around you and identify rotating part around you. In the mechanical industry also there are many rotating parts, for example, rotating shafts, axles, drills, gears and so on. It becomes extremely important to control these rotating parts such that they rotate without boggling and generates minimum vibrations. It becomes important to increase the life cycle of rotating parts, connecting parts and also better safety.

§ Identification of datum feature and datum axis §

For tolerance of rotating parts:
- We must identify the axis of rotation which we call datum axis.
- The datum axis must be derived from a **related functional** part which we call datum feature. Here related means the

datum feature and feature under consideration are connected. Functional means the datum feature is a working component.

- Datum axis can be:
  - o derived from another cylindrical feature of long size.
  - o collectively derived from two cylindrical features sufficiently distance apart.
  - o Derived from planar datum feature which is perpendicular to the axis.

Look at Figure 12.1, given below, to understand datum feature and datum axis and surface around the datum axis to be controlled by runout.

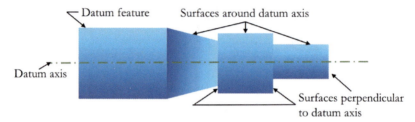

Figure 12.1 – Datum feature, datum axis, and surface for runout

12.1 IDENTIFYING DATUM FEATURES AND AXIS

We begin the process of defining runout tolerances with the identification of datum feature in the entire component which will be used to derive datum axis. Datum feature should be sufficiently long and therefore we selected leftmost feature in Figure 12.1 as the datum feature. Automatically axis of the datum feature will define datum axis, as shown in the figure. Remaining all surfaces, either parallel, inclined or perpendicular to datum axis can be controlled by RUNOUT.

§ **Two cylindrical datum feature** §

If one cylindrical datum feature is not large above, then it may be difficult to identify the datum axis. In such case, we may find two short datums features and combine them to get the datum axis. Look at Figure 12.2, given below, in which two short cylinders at both the ends are selected as combined datum features to derive datum axis.

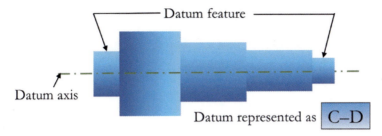

Figure 12.2 – Two datum features to derive datum axis

Datums derived with the help of two cylindrical profiles are represented as C–D as shown in Figure 12.2.

§ One cylindrical one perpendicular surface datum feature §

Look at Figure 12.3 in which neither single long cylindrical datum feature is available nor two short cylindrical datum features are available to define datum axis. In this case, we can select leftmost cylinder as first datum feature and a surface perpendicular to the axis of first datum feature as second datum feature, as shown in the figure. Based on these two datum features, final datum axis can be derived. Such datums are represented in a regular manner showing both datums.

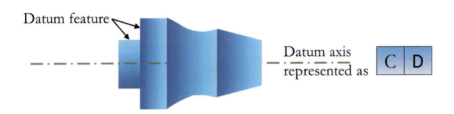

Figure 12.3 – Two datum features to derive datum axis

In fact, datum features can be any feature or combination which is convenient to derive the datum axis during checking the tolerances.

Now we will see how circular runout controls circular profile and total runout controls cylindrical profile of a rotating cylinder.

12.2 CIRCULAR RUNOUT TOLERANCE

Circular runout is used to control the circular profile of a surface. The circular profile can be located at a particular cross-section of a shaft or over a small span of the surface. Look at Figure 12.4, given below, showing four cases in which circular runout can be used.

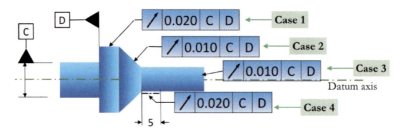

Figure 12.4 – Examples of circular runout

Here one cylindrical feature (on left) and one plane feature are used to define datum axis. **Case 1** is used for small span surface circular runout profiling. It means the circular profile will be checked only at one point on this span and if tolerance comes below 0.020 mm then it will be accepted. **Case 2** is used for the inclined surface. Here again, the profile is checked only at one point. **Case 3** is related to the flat surface perpendicular to a datum axis. **Case 4** is a little different. If you want to control circular runout profile only over a limited span of rotating element, then you use the thick chain to show your area to control the tolerance.

§ **MMC/LMC for runout** §
Since it is used to control the final surface, material condition is not relevant and you did not find it in the example.

§ **Writing runout tolerance** §
The examples we saw above are sufficient.

§ **Inclusion of other tolerances** §
By using this tolerance, we can control the position of the centre of circular profile and circularity of surface profile with reference to a datum axis.

§ **Analysis output of runout tolerance** §

Below is the analysis output for runout tolerance.

Geometric Tolerance	Symbol	Usage	Datum	MMC LMC	Inclusion
Runout	⟋	2D circular profile	Yes	No	Position Circularity

Figure 12.5 – Analysis output of runout tolerance

12.3 TOTAL RUNOUT TOLERANCE

Total runout is used to control circular elements of all surface elements of a cylindrical surface. Look at Figure 12.6, given below:

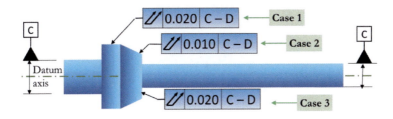

Figure 12.6 – Examples of total runout

Here, you can notice that the chain line is removed. It means we need to check the circular profile throughout the surface.

§ **MMC/LMC for total runout** §

Since it is based on the mass distribution of feature and datum feature, material condition is not used.

§ **Writing total runout tolerance** §

The examples we saw above are sufficient.

§ **Inclusion of other tolerances** §

It is a very useful and easy to use tolerance control. It includes concentricity, perpendicularity, cylindricity, circularity, straightness and circular runout.

§ **Analysis output of total runout tolerance** §

Below is the analysis output for symmetry tolerance.

Geometric Tolerance	Symbol	Usage	Datum	MMC LMC	Inclusion
Total runout	⟋⟋	3D circular profile	Yes	No	Concentricity Perpendicularity Cylindricity, Circularity, Straightness Circular runout

Figure 12.7 – Analysis output of runout tolerance

Chapter recap

In this chapter, we learnt about two geometric tolerances as part of RUNOUT tolerance type namely circular runout and total runout. Below are the main points covered:

- These are primarily used to control boggling of rotating parts
- Identifying datum(s) is important as it has to be a part of the component itself to ensure no boggling. Datums can be cylindrical rotating part or a plane perpendicular to the axis of rotation.
- If circular profile control is needed only at a circle, then we use circular runout and if circular profile control is needed over entire surface then we use total runout.
- Total runout is very useful as it includes much other controls.
- We consolidated our learning as given in below form:

Geometric Tolerance	Symbol	Usage	Datum	MMC LMC	Inclusion
Runout	↗	2D circular profile	Yes	No	Position Circularity
Total runout	⟋⟋	3D circular profile	Yes	No	Concentricity Perpendicularity Cylindricity, Circularity, Straightness Circular runout

If any point is unclear, then clarify it before you proceed.

13 MULTIPLE FEATURE CONTROL TOLERANCES

As a designer, you may like to use multiple feature controls to get the desired output. For example, you may like to control the position and perpendicularity of a hole to ensure proper fit. There are multiple such conditions which are broadly divided into three parts:

13.1 MULTIPLE FEATURE CONTROL

This is used when we need to provide more than one geometric tolerance to a feature. For example, we may need to provide position and perpendicularity tolerance of a hole. It will be written as shown in figure 13.1, given below.

Figure 13.1 – Example of multiple feature controls

Word of caution: If any tolerance is inclusive of other tolerance then you don't need to mention the included one unless you want to control it differently. Look at Figure 13.2 in which straightness and flatness are used. Since straightness is included in flatness, as we learnt in section 8.2, so we don't need to mention straightness with flatness when tolerance values are same. But we need to mention both when tolerance values and feature control methods are different. Therefore, straightness on the left side is not required as tolerance values are same but straightness is required in right side case where tolerance values and feature controls are different.

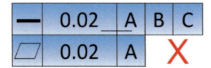 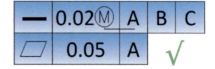

Figure 13.2–Inclusiveness examples for multiple feature control

13.2 COMPOSITE FEATURE CONTROL

This is used when we need to provide single tolerance but different values in different conditions. Look at Figure 13.3, given below. The picture on left side show profile of a surface having a tolerance of 0.010 when considered only with datum A but tolerance is increased when you control it with respect to three datums (A, B, and C) at a time. More control means tolerance has to be relaxed to keep it practical. When you look at the picture on the right side then you would find position tolerance to be 0.02 when considered with datums A and B with any material condition. But when datum feature A is considered with MMC when tolerance has been increased to 0.05.

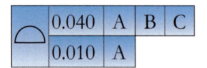 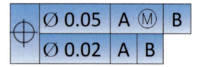

Figure 13.3 – Composite feature controls examples

13.3 COMBINED FEATURE CONTROL

When a feature, controlled by geometric tolerance, also serves as a datum feature then you get a combined feature control, as shown in Figure 13.4.

Figure 13.4 – Combined feature controls examples

Above are just a few examples. You can combine them in any manner you want. Just because this chapter is short, we are skipping the revision of the chapter.

14 BOUNDARY AND ENVELOPE

14.1 CONCEPTS OF A BOUNDARY

As a designer, you know your requirements and maximum deviations you can allow, during the manufacturing of the components, to keep your parts functional. These deviations are caused by:

- Dimensional tolerances (10 ± 1, $60^0 \pm 5^0$, etc.)
- Geometric tolerances (position, straightness, etc.)
- Bonus tolerances (departure from MMC/LMC)

As a designer, you are concerned about the combined effect of all of the above deviations. You would calculate the combined effect of all tolerances to find out maximum possible deviations. These maximum deviations are called **boundaries**. Look at Figure 14.1, given below:

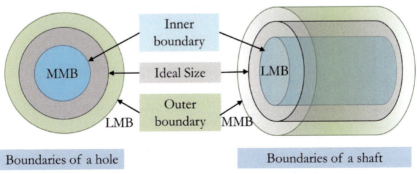

Figure 14.1 – Inner, outer boundaries (MMB, LMB)

In Figure 14.1, we are taking examples of the hole (to represent all internal features) and shaft (to represent all external features). Grey features are ideal size features, blue are smallest size features and green features are largest size features.

Smallest features are called "**Inner Boundary**" and largest features are called "**Outer Boundary**". Let's understand the meaning of each of them and also the formula to calculate their values.

We will start with **inner boundaries (IB)**:

- The smallest possible hole is a condition due to MMC of the hole and the further reduction in hole size due to all geometric tolerances (GT) and all bonus tolerances (BT). So, we can say:
$$IB_{Hole} = MMC_{Hole} - \sum(GT) - \sum(BT)$$
This is the maximum material boundary (**MMB**) of a hole.

- Similarly, the smallest possible shaft is a condition due to LMC of the shaft and the further reduction in size due to GT and BT. So:
$$IB_{Shaft} = LMC_{Shaft} - \sum(GT) - \sum(BT)$$
This is the least material boundary (**LMB**) of a shaft.

Now we will pick up **outer boundaries (OB)**:

- The largest possible hole is a condition of LMC of the hole and a further increase in size due to GT and BT. Here we can say:
$$OB_{Hole} = LMC_{Hole} + \sum(GT) + \sum(BT)$$
(Notice the change in sign as highlighted by yellow colour. It is due to the addition of sizes to move from LMC to bigger size boundary)
This is the least material boundary (**LMB**) of a hole.

- Similarly, the largest possible shaft is a condition due to MMC of the shaft and increase in size due to GT and BT. So we can say:
$$OB_{Shaft} = MMC_{Shaft} + \sum(GT) + \sum(BT)$$
This is the maximum material boundary (**MMB**) of a shaft.

Without any stress, we learnt about **"Maximum Material Boundary"** (MMB) and **"Least Material Boundary"** (LMB). Actually, we started with simple inner and outer boundaries concept and found the relationship between inner and outer boundaries with MMB/LMB as summarized in Figure 14.2, given below:

Boundary	Hole (internal features)	Shaft (External features)
Inner boundary (Smallest)	MMB = MMC-∑GT-∑BT	LMB = LMC-∑GT-∑BT
Outer boundary (Largest)	LMB = LMC+∑GT+∑BT	MMB = MMC+∑GT+∑RT

Figure 14.2 – Inner, outer boundaries, MMB, and LMB

Below are definitions of LMB and RMB from ASMEY14.5-2009:

LMB: the limit defined by a tolerance or combination of tolerances that exists on or inside the material of a feature(s).
MMB: the limit defined by a tolerance or combination of tolerances that exists on or outside the material of a feature(s).

You may validate your learning by checking for LMB which is said to be of the material. It is shown in Figure 14.2 as inner boundary of the shaft which will be inside the shaft and the outer boundary of the hole which will again be inside the hole. Similarly, you can validate RMS cases as well.

There is one more term, RMB (regardless material boundary)

RMB is derived from actual parts by moving from MMB to LMB and stopping when contact is maximum. It is done with the help of theoretical simulator or practical simulators. Look at Figure 14.3 explaining the RMB concept on actual part.

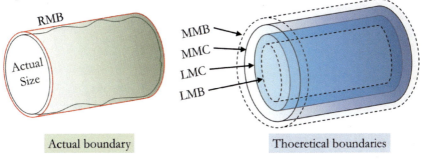

Actual boundary   Thoeretical boundaries

Figure 14.3 – Showing RMB with MMB and LMB

One very important interpretation is given in Figure 14.3, that is, MMC and MMB are not same, and similarly LMC and LMB are not same. For shafts, MMB with being larger than MMC and LMB will be smaller than LMC. So you should not confuse them. They all are different. Isn't it?

 Question time !

Can you think the situations when theoretical MMB will be the same as actual MMB and LMB will be the same as actual LMB?

Answer:
The MMB would be an actual MMB if the tolerance (location or orientation) for that datum feature was zero at MMC.
The LMB would be an actual LMB if the tolerance (location or orientation) for that datum feature was zero at LMC

14.2 VIRTUAL CONDITION AND RESULTANT CONDITION BOUNDARIES

These are boundaries defined by dimensional tolerance and geometric tolerance when material conditions are applied to the feature of size. Let us look at the definition of these terms:

Virtual Condition
A constant boundary generated by the collective effects of a considered feature of the size's specified MMC or LMC and the geometric tolerance for that material condition.

Resultant Condition
The single worst-case boundary generated by the collective effects of a feature of the size's specified MMC or LMC, the geometric tolerance for that material condition, the size tolerance, and the additional geometric tolerance derived from the feature's departure from its specified material condition.

In order to simplify, we can say:
Virtual condition boundary is a dimensional boundary at given material condition adjusted by geometric tolerance crossing the given material condition.

Resultant condition boundary is a dimensional boundary opposite to given material condition adjusted by geometric tolerance crossing opposite material condition (LMC is opposite of MMC and vice versa) and additional maximum bonus tolerance. *Maximum bonus tolerance is the difference between MMC and LMC (learnt earlier).*

Let's take a few examples to confirm our understanding.

Example 1: Internal feature (hole) with GT at MMC
Given, Dimension: 30.1 to 30.5, GT: 0.1 positional @ MMC
We learnt, BT will vary from 0 (at MMC) to (30.5-30.1=0.4 (at LMC)
Therefore, $BT_{max} = 0.4$
Virtual condition boundary
= Size at MMC $\pm$ GT crossing MMC
= 30.1 + (-0.1) = 30.0
Resultant condition boundary
= Size at LMC $\pm$ (GT crossing LMC + BT_{max})
= 30.5 + 0.1 + 0.4 = 31.0

Example 2: Internal feature (hole) with GT at LMC
Given: Dimension: 30.1 to 30.5, GT: 0.1 position @ LMC
$BT_{max}=0.4$ (at MMC)
Virtual condition boundary
= Size at LMC $\pm$ GT crossing LMC
= 30.5 + (0.1) = 30.6
Resultant condition boundary
= Size at MMC $\pm$ (GT crossing MMC + BT_{max})
= 30.1 – (0.1 + 0.4) = 29.6

Example 3: External feature (shaft) with GT at MMC
Given, Dimension: 29.5 to 29.9 and GT: 0.1 position @ MMC
$BT_{max} = 0.4$
Virtual condition boundary
= Size at MMC $\pm$ GT crossing MMC
= 29.9 + (0.1) = 30.0
Resultant condition boundary
= Size at LMC $\pm$ (GT crossing LMC + BT_{max})
= 29.5 – (0.1 + 0.4) = 29.0

Example 4: External feature (shaft) with GT at LMC

Given, Dimension: 29.5 to 29.9 and GT: 0.1 position @ LMC

$BT_{max} = 0.4$

Virtual condition boundary

= Size at LMC $\pm$ GT crossing LMC

= 29.5 + (-0.1) = 29.4

Resultant condition boundary

= Size at MMC $\pm$ (GT crossing MMC + BT_{max})

= 29.9 + (0.1 + 0.4) = 30.4

❋ Your thought bite ❋

You got the essence of the virtual condition and resultant condition boundaries. Can you calculate these at RFS for given conditions?

Hey! Wait. You got into the trap. You cannot compute it.

I mentioned at the beginning of this section, these values are applicable only to features of size with the given material condition. So these concepts do not apply to RFS. However same learning is used to calculate inner and outer boundaries for RFS features by simplified formula as BT is always zero for RFS (because there is no MMC/LMC condition applied). Let's take same hole and shaft examples to find out inner and outer boundaries.

Example 5: Internal feature (hole) with GT (RFS)

Given, Dimension: 30.1 to 30.5, GT: 0.1 positional (RFS)

Here, $BT_{max} = 0.0$

Inner boundary

= Size at MMC $\pm$ GT crossing MMC

= 30.1 + (-0.1) = 30.0

Resultant condition boundary

= Size at LMC $\pm$ (GT crossing LMC + BT_{max})

= 30.5 + 0.1 + 0.0= 30.6

Example 6: External feature (shaft) with GT (RFS)

Given, Dimension: 29.5 to 29.9 and GT: 0.1 position @ LMC

Here, $BT_{max} = 0.0$

Inner boundary

= Size at LMC $\pm$ GT crossing LMC

= 29.5 + (-0.1) = 29.4
Resultant condition boundary
= Size at MMC $\pm$ (GT crossing MMC + BT_{max})
= 29.9 + (0.1 + 0.0) = 30.0

Now we can say, our learning is simplified, yet perfect! Yay! Our simplified concept for virtual and resultant conditions are:
Virtual condition = dimensional boundary at given material condition adjusted by geometric tolerance crossing the given material condition.
Resultant condition boundary = dimensional boundary at opposite to given material condition adjusted by geometric tolerance crossing opposite material condition and additional maximum bonus tolerance.

 Question time !

Question 1: A hole has dimension 20.0-20.6 with a straightness tolerance of 0.1 at MMC. Find:
   A. Value of virtual condition of the feature
   B. Bonus tolerance of the feature if actual size is 20.6
   C. Bonus tolerance of the feature if actual size is 20.4

Question 2: A shaft has dimension 10.0-10.5 with a straightness tolerance of 0.2 at MMC. Find:
   A. Value of virtual condition of the feature.
   B. Bonus tolerance of the feature if actual size is 10.5.
   C. Bonus tolerance of the feature if actual size is 10.1.

Answers:
1.A – 19.9; 1.B – 0.6; 1.C – 0.4
2.A – 10.7; 2.B – 0.0; 2.C – 0.4

14.3 CONCEPTS OF ENVELOPE

Actual mating envelope

Envelope concept is used to validate the actual output. In simple terms, **it is the closest fit counterpart perfect feature**. For holes, an envelope will be the largest size shaft which can establish maximum contacts with hole surface. Similarly, for shafts, an envelope will be a hole of smallest size with maximum contacts.

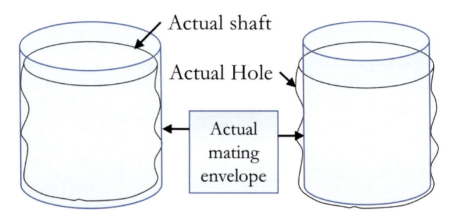

Figure 14.4 – Illustration of Actual mating envelopes

Look at Figure 14.4, given above. On the left side, you see an irregular shape representing actual shaft. On top of it, there is an envelope which is closest to the shaft and also it is perfect in shape. On the right side, you find an irregular hole which has one perfect shaped closest fit shaft. These two mating shapes are mentioned as the **closest fit counterpart perfect features** in the previous paragraph. These are examples of actual mating envelopes.

There are two types of envelopes:
1. Unrelated actual mating envelope: It is closest fit counterpart perfect feature without considering any datum feature or constraint.
2. Related actual mating envelope: It is also the closest fit counterpart perfect feature while constrained either orientation or location, or both by applicable datum(s).

Look at Figure 14.5, given below, to understand the difference.

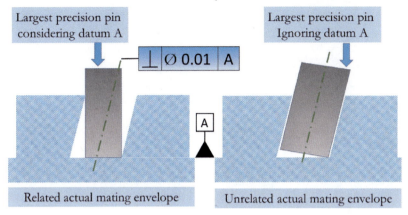

Largest precision pin considering datum A

⊥ | Ø 0.01 | A

Largest precision pin Ignoring datum A

A

Related actual mating envelope

Unrelated actual mating envelope

Figure 14.5 – Related and unrelated actual mating envelope

Suppose you gave a requirement to make a hole with perpendicularity tolerance of 0.01 related to datum A. The hole was made as shown above. Now you want to find actual mating envelop with the help of a precision pin. When you consider the datum condition when you put a precision pin perpendicular to datum A, then you will be able to insert thinner pin as shown in the left side picture. On another hand when you ignore perpendicularity requirement, then you can insert thicker precision pin as shown in the picture on the right side.

Size of the precision pin on the left side gives you related actual mating envelope and size of the precision pin on the right side gives you an unrelated actual mating envelope. Logic is simple, one is related to datum and other is unrelated to the datum, hence the name was given accordingly.

14.4 FUNDAMENTAL RULES OF GD&T

Rule #1: Perfect Form at MMC – Individual Feature of Size rule
Where only a tolerance of size is specified, the limits of the size of an individual feature prescribe the extent to which variations in its form—as well as in its size—are allowed.
In simple terms, *dimension tolerance should never cross-boundary conditions when geometric tolerance is not provided.*

Example 1: If a shaft dimension to be Ø9.5 – 10.0 mm, then shaft must pass through a perfect hole of Ø10.0 mm as shown in Figure 14.6:

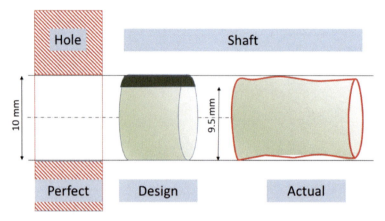

Figure 14.6 – Illustration for Rule # 1 example 1

Example 2: If a hole dimension to be Ø12.0 – 12.5 then a Ø12.0 mm perfect pin must pass through the hole as shown in Figure 14.7:

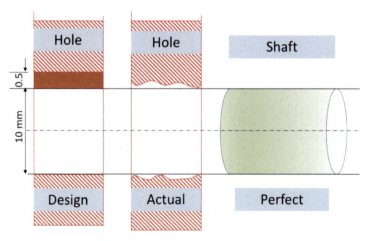

Figure 14.7 – Illustration for Rule # 1 example 2

An important point from Rule #1:

There cannot be any form error (like circularity, straightness, etc.) at MMC. means, amount of form error is less than or equal to deviation from MMC.

Rule #2 – The All applicable Geometric Tolerances Rule

RFS applies, with respect to individual tolerance, datum reference, or both, where no modifying symbol is specified. MMC or LMC must be specified on the drawing where required.

This is really simple to understand. In short, RFS is the default material condition for tolerance if MMC or LMC is not applicable.

Chapter **recap**

In this chapter, we learnt about:

- Concepts of Boundary
    - o   Inner boundary
    - o   Outer boundary
    - o   Maximum material boundary (MMB)
    - o   Least material boundary (LMB)
    - o   Regardless feature boundary (RMB)
- Concepts of the actual mating envelope
    - o   Related actual mating envelope
    - o   Unrelated actual mating envelope
- Fundamental rules of GD&T
    - o   Rule # 1
    - o   Rule # 2

If any point is unclear, then clarify it before you proceed.

15 IMPORTANT MODIFIERS IN GD&T

Like MMC/LMC modifiers, there are few more modifiers which add a description to dimension or geometric tolerance to communicate more details. In this section, we will cover multiple such modifiers.

15.1 Translation

It is used when the better fit is given more importance than the position of the fit. Suppose we are designing a mechanism to measure the rotational speed of a shaft. We decided to fit an extrusion on the surface which will mark each rotation. Figure 15.1 given below shows shaft in blue colour, extrusion is shown in golden colour and datum of extrusion central plane or slot in transparent is shown in red colour.

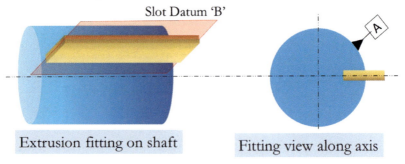

Slot Datum 'B'

Extrusion fitting on shaft

Fitting view along axis

Figure 15.1 – a Use case for translation modifier

In Figure 15.2, we can see datum B has been allowed translation. It means better fitting is expected even if datum B needs to be translated because even after translation rotation can be captured.

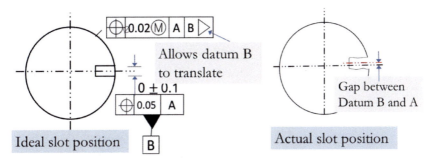

Ideal slot position

Actual slot position

Figure 15.2 – GD&T representation of translation modifier

The right side of Figure 15.2 shows slot after first machining. Now refer to Figure 15.3 in which fitting is found to be perfect but the position is shifted. If position HAS TO BE corrected then rework is needed to widen the slot but it will result in loose-fitting, as shown in the right side of Figure 15.3. Obviously, if shifting of extrusion is allowed then we want a situation in the left figure. This is the use of Translation modifier.

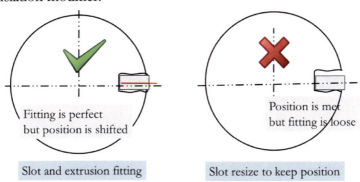

Slot and extrusion fitting

Slot resize to keep position

Figure 15.3 – Use of translation modifier

As a conclusion, we can say "**Translation not only helps in reducing rejections but also provided better fitting**". It becomes very significant for rotational or rocking fitting where loose contacts fail very fast. You know what to do now!

15.2 Projected tolerance zone

It is used for projecting (or extending) the tolerance for continued mating. For example, continuity of threads in the perpendicular direction. It is generally useful for positional or orientation tolerances for desire fit between holes and fasteners, like screws, studs, or pins. Let's take an example of the thread profile to understand the benefit of the projected tolerance zone

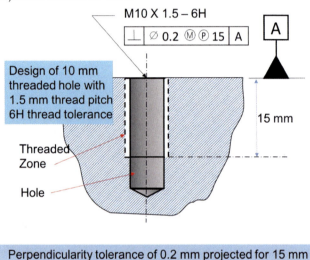

Perpendicularity tolerance of 0.2 mm projected for 15 mm

Figure 15.4 – Projected tolerance zone example

Figure 15.4 shows a design of a threaded hole with a 15 mm thread depth. If the functional requirement is to limit the perpendicularity of thread hole to 0.2 mm diameter, then we will provide a projected tolerance as 15 mm as shown in the figure.

Look at Figure 15.5. Picture of the left side shows the impact of providing tolerance at MMC, in terms of maximum tolerance. The picture on the right side shows a case when we do not provide projected tolerance. The only constraint is to keep the centre of the hole in 0.2 mm tolerance area. Depth can vary in any direction, making effective projected tolerance zone much larger. Thanks, GD&T, to provide projected tolerance!

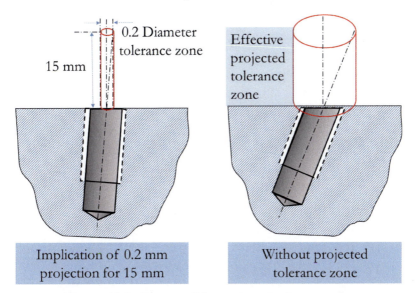

Figure 15.5 – Impact of the projected tolerance zone

15.3 Free state   Ⓕ

Non-rigid parts (for example, thin wall vessel) may change shape after removal of forces applied during manufacture, due to its weight and flexibility, and release of internal stresses developed during fabrication. Free state of the non-rigid part is achieved after all forces are removed. We may need the part to meet its tolerance requirements while in the free state.

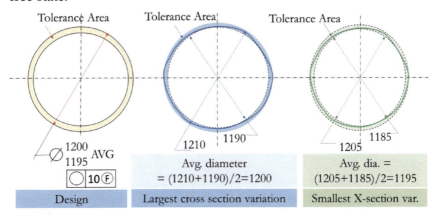

Figure 15.6 – Illustration of free state modifier

In Figure 15.7, you may see the tolerance requirement in a free state (with symbol 'F' in a circle) on the leftmost picture. Pictures in middle and right side show allowed variations at different cross-sections. The middle picture shows largest allowed cross-section variation and the rightmost picture shows smallest allowed cross-section variation.

15.4 Tangent plane ⓣ

It is generally used when we place another object on the surface, Here tangent plane established by the contacting points of a surface becomes important, for example, to place another part, the tangent plane symbol is added. Let's consider a platform with the horizontal top on which another object will be placed as shown in Figure 15.7.

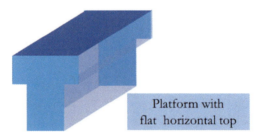

Platform with
flat horizontal top

Figure 15.7 – Example model for tangent modifier

The tabletop needs to be horizontal with max 0.1 mm level difference. How would you communicate it? Flatness? No, it's only for variation between two planes. Parallelism? Can be, but not ideal. See Figure 15.8.

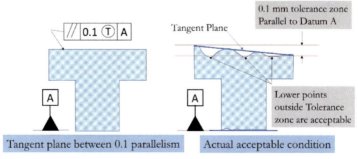

Figure 15.8 – Illustration for tangent modifier

Here, the actual surface goes beyond tolerance zone but it is acceptable as long as a tangent plane in this region remains in 0.1 mm tolerance zone, as shown on right side picture. Do notice the profiling on the bottom which creates datum feature A. It makes a difference to have minimum tolerance in datum feature.

15.5 Unequally disposed profile Ⓤ

This modifier is used to control the surface profile tolerance zone across the ideal surface profile. Let's take the same example of a car hood. Using this modifier, you can distribute total tolerance above and below the ideal profile of hood to control allowed variation on a particular side which is more critical. For example, the top surface of car hood needs to look better as compared to the bottom part of car hood because the top part is always visible and the bottom part is seen only when you open the hood for maintenance activities. Look at Figure 15.9 in which car hood surface profile is being provided.

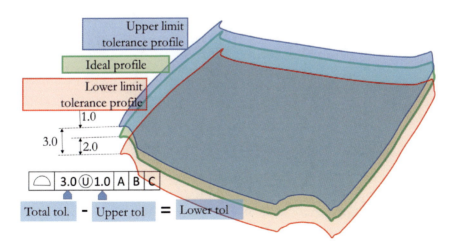

Figure 15.9 – Illustration for unequally disposed of modifier

Here green profile, in between blue and red profiles, represents ideal hood profile, upper blue profile represents maximum deviation on the upper side and red profile represents maximum deviation on the lower side of the car hood. Look at the GD&T details. Here designer wants to have minimum deviations on the top surface and has provided a total deviation of 3.0 mm but allowed only 1.0 mm deviations for upper profile and remaining 2.0 mm to be used for a lower profile. Altogether 3.0mm tolerance is given but by using this modifier designer has attempted to control a better finish of the top surface.

Sometimes, the upper deviation is called, the *maximum addition* of material on idea profile and lower deviation is called the *maximum removal* of material on the ideal profile. It makes more sense when you try to finish the upper layer of a solid body by means of adding or removal of materials.

15.6 Independency ⓘ

GD&T rule # 1 says, there cannot be any form error (like circularity, straightness, etc.) at MMC, means form would be perfect at MMC. It may not be necessary to achieve perfection at MMC (or LMC) to meet functionality. In such cases, this modifier can be used as shown in Figure 15.10, given below:

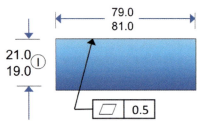

Figure 15.10 – Illustration for independency modifier

Here putting independency means the feature can be controlled independently without considering any datum. For example, making finished tabletop with specified dimensions of thickness and flatness only on top. This is very useful tolerance relaxation for mass production to reduce cost.

15.7 Statistical tolerance

This is related to cumulative tolerance of an assembly, called tolerance stack. There are two types of tolerance stack analysis:

A. *"Worst Case Analysis"*. In this approach, we simply add all linear tolerances in one direction. Suppose you have 5 linear parts which join together to form an assembly. If all have a linear tolerance of 1.0 mm, then in worst-case analysis, total tolerance of assembly would be 5.0 mm. This approach is extremely safe but costly to achieve desired assembly tolerance.

B. *"Statistical Tolerance Analysis"*: Any manufacturing has deviation among all parts produced. It can safely be assumed to follow a normal distribution as shown in Figure 15.11, given below.

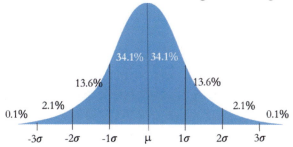

Figure 15.11 – Normal distribution graph

The output of this analysis says *the assembly tolerance is equal to the square root of the sum of the squares of the individual tolerances*. Based on this rule, assembly tolerance in the same case would be $\sqrt{(1^2+1^2+1^2+1^2+1^2)} = \sqrt{5}$. Figure 15.12 is an illustration of assigning statistical tolerance.

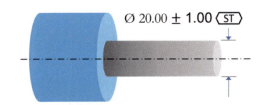

$\varnothing\ 20.00 \pm 1.00$ ST

Figure 15.12 – Illustration of statistical tolerance

By applying statistical tolerancing, tolerances of individual components may be increased or clearance between mating parts may be reduced. The increased tolerance or improved fits may reduce manufacturing cost or improve the product performance, but should only be employed where appropriate statistical procedure control is used.

15.8 Continuous feature 〈CF〉

This modifier is very useful for the repeated feature, or continuation of a feature across slots, as shown in Figure 15.13, given below.

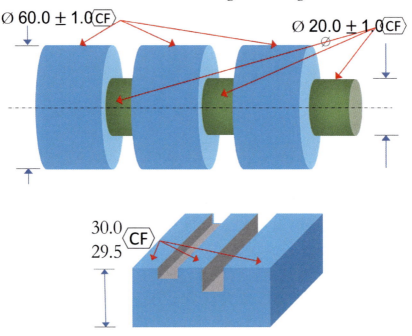

Figure 15.13 – Illustration of continuous modifier

15.9 Controlled radius   CR

This modifier is used when the circular profile has to have a fair/smooth curve with radius at all points to remain within given radii limits. Refer to Figure 15.14, given below, showing two curves example.

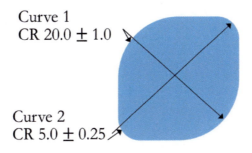

Curve 1
CR 20.0 ± 1.0

Curve 2
CR 5.0 ± 0.25

Figure 15.14 – Use case for controlled radius modifier

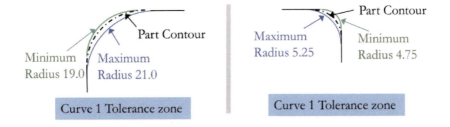

Part Contour

Minimum
Radius 19.0

Maximum
Radius 21.0

Curve 1 Tolerance zone

Part Contour

Maximum
Radius 5.25

Minimum
Radius 4.75

Curve 1 Tolerance zone

Figure 15.15 – Illustration of controlled radius modifier

15.10 Dimension origin

It is used to define a reference plane for dimension measurement, which becomes important in a few cases. One such case is explained below with the help of Figures 15.16 and 15.17:

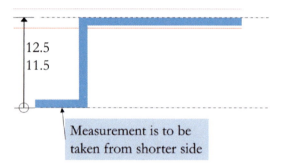

12.5
11.5

Measurement is to be taken from shorter side

Figure 15.16 – a Use case for dimension origin modifier

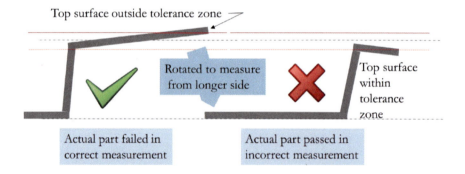

Top surface outside tolerance zone

Rotated to measure from longer side

Top surface within tolerance zone

Actual part failed in correct measurement

Actual part passed in incorrect measurement

Figure 15.17 – Illustration of dimension origin modifier

15.11 Between

This modifier indicates that a tolerance or other specification apply across multiple features or portion(s) of the features as mentioned between symbol, F, and G in the example shown in Figure 15.18. Here, F and G may be points, lines, planes or features

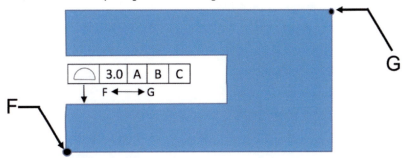

Figure 15.18 – Illustration of between modifier

15.12 All around

This is applied to a profile to extend tolerance all around the profile shown in the drawing of a particular view. Consider the example given below in Figure 15.19. Here, line profile is extended all around by placing the single circle on leader dimension line. When you place it, you don't need to make the same notation on all three line profiles.

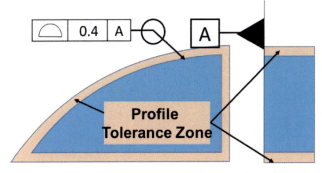

Figure 15.19 – Illustration of all around modifier

15.13 All over

This is applied to a profile to extend tolerance to the entire 3D surface of the part. Consider the example given below in Figure 15.20. Here profile is extended to the entire 3D surface by placing the double circle on leader dimension line. When you place it, you don't need to make the same notation on all profiles.

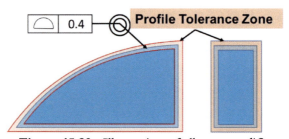

Figure 15.20 – Illustration of all over modifier

15.14 Individually   INDIVIDUALLY

It is a great help for repetitive patterns of features. Look at Figure 15.21 in which one plane has six sets of holes and in each set, you have one large and 4 small holes. It is highly time-consuming if you work on all features explicitly.

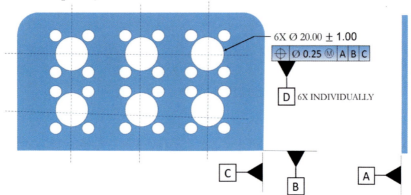

Figure 15.21 – Illustration of all over modifier

In order to make it convenient, the designer has put 6X INDIVIDUALLY which means each bigger hole will work like datum feature D individually for four small holes around them. If individually word was not mentioned then there we would have to derive datum D based on six holes and use the same datum D for next 24 holes, which would be complicated and difficult to make. So individually helps in localized manufacturing over parts with patterns of features.

Chapter  recap

In this chapter, we learnt about menu modifiers which not only helps designers to communicate important information but also increases efficiency. Below are the modifiers learnt:

- Translation
- Projected tolerance zone
- Free state
- Tangent plane
- Unequally disposed profile
- Independency
- Statistical tolerance
- Continuous feature
- Controlled radius
- Dimension origin
- Between
- All around
- All over
- Individually

If any modifier is unclear, then clarify it before you proceed.

16. INDUSTRIAL APPLICATION EXAMPLES

As we all know, there exists a huge gap between theoretical knowledge and practical knowledge. There lies a huge learning when we work on real-life works. Highest level of learning is achieved on the job. And, therefore, the experience is important. Isn't it?

In this chapter, we will try to fast track industrial learning by looking at 20 actual industrial examples, which we thankfully received from a few multinational companies. In order to hide their identity and protect their design details, we will not share their drawing in as is form, rather we would make our own models with features and GD&T applications similar to original component., such that we can extract following important earnings out of them:
- The nature of the features on which GD&T was applied
- Example of GD&T application, in an as-is form used by industry
- Discuss the reasons for GD&T application
- Selection of datum to apply GD&T
- Review of feature condition consideration (MMC/LMC/RFS) made by industry
- Compare the ASME way of suggested standards and actual industrial works.

You are targeted for (minimum) following learning:
- Industry adapts all beneficial points, excludes points which make their work inefficient,
- Corrects the mistakes in standards (if any) by themselves to move ahead and stay ahead of standards.
- Standards are made to help industry based on the learning from the industry such that all are benefited and unified standard can be derived. It is a cyclic process between standardization organizations likes ASME, ISO, etc. and the industry.

To complete our learning process, we will review 14 examples, one each for each geometric tolerance, to provide real-life examples, understand how the industry works, and be ready for the industry. Come along, let's see the actual world of GD&T!

16.1. STRAIGHTNESS INDUSTRIAL EXAMPLE

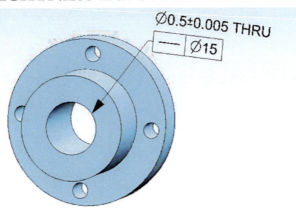

Figure I.1– Example of Straightness

Figure 1.1 shows an example of the straightness of an axis. Notice the following points in GD&T application:

- Matching to our learning, it does not have any Datum.
- We learnt, material condition (MMC/LMC) is applied for axis straightness, however, we don't find it in industrial example. Why? If you are saying, the designer applied RFS for which nothing is to be mentioned then you are perfectly right. Okay, another question, why RFS is applied, why not MMC, which is generally applied on straightness? First hint: *tolerance at MMC or LMC generates bonus tolerance*. Now getting the clue? Yes, the designer did not intend to give any bonus tolerance and therefore RFS was used. You are right, again.

Let me check your thought process. What are other possible geometric tolerances which can be applied on the same hole? Perpendicularity? Position? Circularity? Cylindricity? Probably all of these can be applied, but why would you do so. If only straightness is sufficient to control the feature to remain functional as per need then you should not add more GD&T to avoid increase of complexity and cost. Remember, main intention of the GD&T is to reduce cost, that's the fundamental, that's the GD&T mantra.

16.2. FLATNESS INDUSTRIAL EXAMPLE

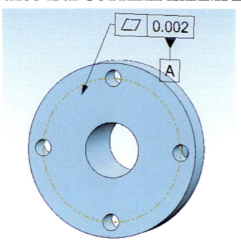

Figure I.2 – Example of Flatness

The above picture shows the rear side of the same part used in previous example. Here the designer has provided flatness to the back surface, in order to fit properly to adjacent part. The same controlled surface has been designated as datum A. Here the flatness is given without any datum, which is again matching with our understanding. Isn't it? So all well for our flatness GD&T learning.

16.3. CIRCULARITY INDUSTRIAL EXAMPLE

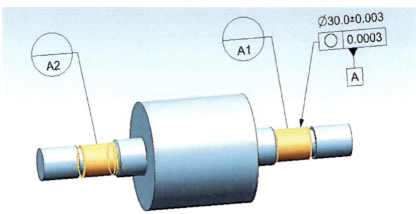

Figure I.3 – Example of circularity

Trust me it was time taking job to arrange for an industrial example of circularity. Why is it so? Don't industry like it? Do they find it useless? Or maybe, instead of circularity, they like to use runout or total runout, which covers circularity? Yes, you are right, the last option is correct one. Industry prefers runout and total runout over circularity. Anyways, let's look at the example shown in Figure I.3,

Well, the situation is a little tricky here, but we will simplify it. Highlighted portions of the part are identified as critical features of the part. Two belt drives are planned to run over these areas. Therefore, the designer decided to control circularity of these two features, and also derive Datum A based on these two features. Look at datum target A1 and A2. These two datum targets are used to derive datum A. So the intention of the designer to control the datum features to have good quality datum is achieve in this process. Wasn't it interesting? This is the beauty of industrial examples. You learn much more than what you may learn in theoretical models and explanations.

16.4. CYLINDRICITY INDUSTRIAL EXAMPLE

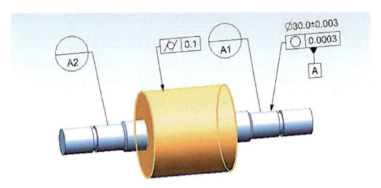

Figure I.4 – Example of Cylindricity

We are referring to the same model used in the previous example. This time the central thick cylindrical feature is our point of focus and has been highlighted in orange colour. Here, the designer wants to control cylindricity, as shown in figure. We don't see Datum reference in this example. It is matching our learning of datum being not required for cylindricity, as it is independent feature to be controlled. So we are becoming GD&T literate.

16.5. PROFILE OF A LINE INDUSTRIAL EXAMPLE

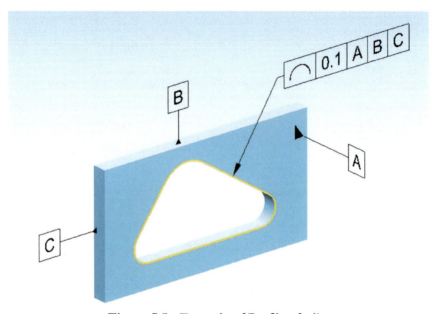

Figure I.5 – Example of Profile of a line

In section 9.1, we saw the illustrative example of the profile of a line with two datums. Theoretically, 2 datums may be sufficient but in above industrial example, the designer has given three datums. Why? Whether the third datum unnecessary? What is the intention of putting it? The answer is simple, The designer suggests to measure the profile while manufacturing and quality check by fixing the part by three constraints aligned to datums A, B, and C.

Exclusive new learning from this industrial example:

Even though two datums may be sufficient to provide Profile os a line geometric tolerance, however as a designer you may like to advice three datums for better feature control.

16.6. PROFILE OF A SURFACE INDUSTRIAL EXAMPLE

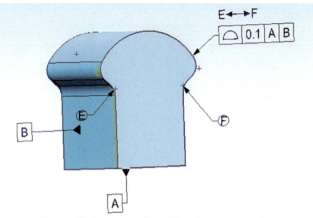

Figure I.6 – Example of Profile of a Surface

Figure I.6, given above, shows a part which looks similar to a bread loaf. The designer intends to control upper profile between points A and B. Our learning for section 9.2 is matching with above GD&T example. So we are good at this geometric tolerance.

16.7. ANGULARITY INDUSTRIAL EXAMPLE

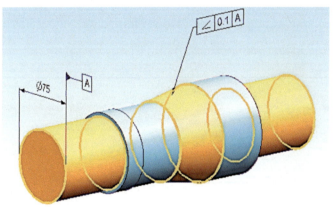

Figure I.7 – Example of Angularity

Figure I.7, given above, shows a central part which is angular to the axis of the complete part, defined as datum A. So it matching our learnings. We are good in angularity.

16.8. PERPENDICULARITY INDUSTRIAL EXAMPLE

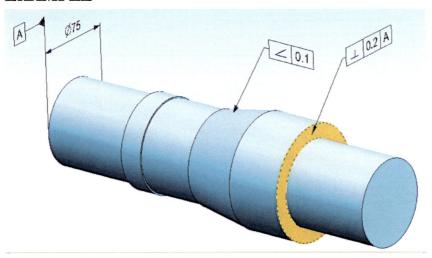

Figure I.8a – Example of perpendicularity of vertical face to an axis

In part shown above, the axis of the part is identified at Datum feature A. The perpendicularity of vertical faces needs only one datum and tolerance should be a simple number, as we learnt. Above industrial example also reflects same convention used. So we are good again!

Now look at Figure I.8b, given below. Here, the designer has identified earlier controlled feature as datum feature B.

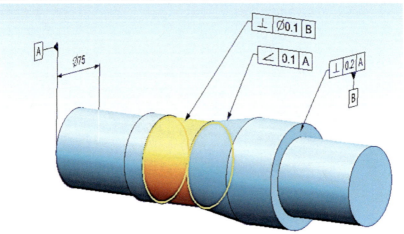

Figure I.8b – Example of perpendicularity of cylindrical profile to a plane

Did you notice symbol Ø for perpendicularity? In section 10.2, when we were learning the basics of perpendicularity in section 10.2 then we never found use of symbol Ø. First question. Is it correct to use Ø symbol in above example? Second question. If yes then what does it represent? Answer is: Ø represents circular geometric tolerance zone applied on the axis of the cylindrical profile. It is justified if the designer intends to provide perpendicularity to the axis. If you want to control cylindrical profile then we would use cylindricity. Isn't it?

Exclusive new learning from this industrial example:

In addition to flat tolerance zone for perpendicularity, it is possible to have cylindrical tolerance zone for cylindrical perpendicular profiles, in consideration with axis of the cylindrical profile.

16.9. PARALLELISM INDUSTRIAL EXAMPLE

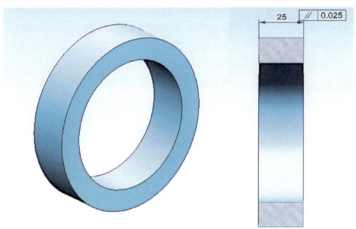

Figure I.9 – Example of parallelism

In section 10.2, we learnt about the parallelism between two planes, taking one of the planes as datum. In industrial example shown in Figure I.9, given above, we find a ring-like part whose both the sides needs to be parallel to each other. Surprisingly, the designer has not used any datum. Did she/he miss it? Not really. It was not needed. The designer has left the discretion to choose datum whichever way the manufacturer wants. The designer's interest is only to ensure parallelism between both sides.

16.10. TRUE POSITION INDUSTRIAL EXAMPLE

Look at Figure I.10 showing a flange with 4 holes at four equally spaced locations with 0.025 position tolerance of circular zone.

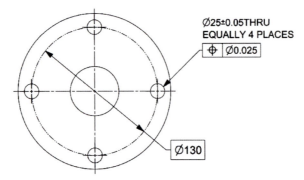

Figure I.10 – Industrial example of Position tolerance

Notice the following points:

- No material condition is given, means it is RFS, means the designer is not providing any bonus tolerance to manufacturer these holes.
- Location (diameter of the circle on which holes are located has been assigned a basic (boxed) dimension.
- We learnt earlier, DATUM to be mandatory for position tolerance. Correct? But we don't find any such datum mentioned here. Does it mean this industrial example is **incorrect** and would lead to miss-communication between designer and manufacturer? The answer is **No**. The designer could have identified the axis of the flange, passing through the center, to be used as datum A and could have referred to same as datum A in positional tolerance. It would have provided the same information which the designer has communicated by providing basic (or boxed) dimension for the location. So there is no information loss. **Yes**, as a standard practice, the designer should have given datum reference in the feature control frame. But industry likes to keep it simple.

Exclusive new learning from this industrial example:

Mandatory datum for positional tolerance can be avoided or replaced by providing location details as basic dimension.

Simplified GD&T

16.11. CONCENTRIC TOLERANCE APPLICATION

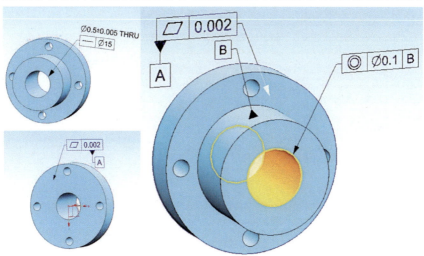

Figure I.11 – Example of concentricity

We take the same model used in first (straightness) and second (flatness) examples. This time the designer has put concentricity of central hole with axis of outer surface, identified by Datum B. Here all points are going as per our learning and standard, means we find symbol Ø to represent cylindrical tolerance zone. and inclusion of datum A in feature control frame. So all well here.

Okay, let me ask a question here.

The designer has another option to provide perpendicularity of central hole axis to datum A, which was already defined, then why the designer did not use it, instead of defining one more datum and then use concentricity tolerance?

Answer: Both the options are possible to use. As a designer, we will select the geometric tolerance which is functionally important. Here designer's requirement is for the concentricity, not the perpendicularity. If both controls are required then designer will apply both conditions.

139

16.12. SYMMETRY INDUSTRIAL EXAMPLE

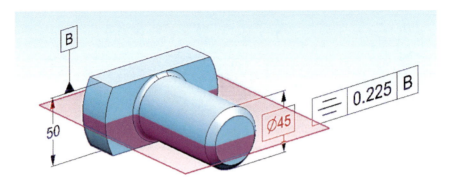

Figure I.12 – Example of symmetry

Figure I.12, given above, shows an industrial example of symmetry across an imaginary pink plane identified as datum B. Here our learning is exactly matching with the industrial example. So we are well on track.

16.13. RUNOUT INDUSTRIAL EXAMPLE

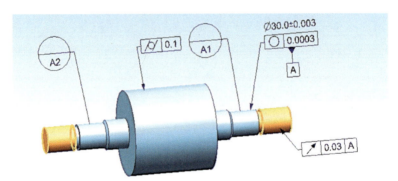

Figure I.13 – Example of Runout

Figure I.13, given above, shows an industrial example of runout. We have taken the same model used earlier to explain the concept. Here runout is applied on features highlighted with orange colour. The application is simple, aligned to our learning. So we are good for runout as well.

16.14. TOTAL RUNOUT INDUSTRIAL EXAMPLE

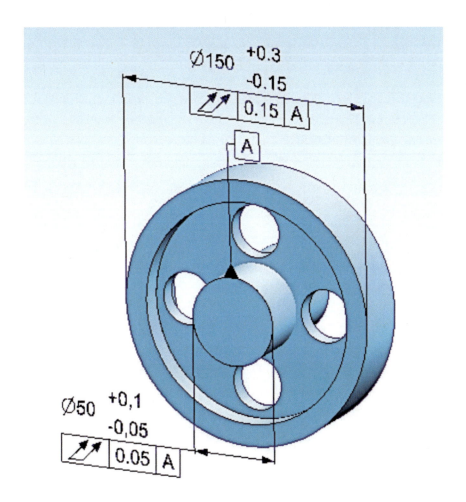

Figure I.14 – Example of Total runout

Let's consider a flywheel design shown above in Figure I.14. The central axis is defined as datum A. We can find total runout to be applied at two places. One is at the central axis and another is at the rim of the flywheel. Notice the values of total runout provided by designer. Can we reverse the values? Actually no, if you want to sound like a sensible designer, because as your distance from rotating axis increases, we would have to provide higher tolerance for total runout. This is the last learning from our industrial examples.

APPENDIX A1: DETERMINING DIMENSIONAL TOLERANCES USING ISO-286

We are referring to ISO 286-1:2010 and ISO 286-2:2010 in particular for the definition of IT Grades.

Tolerances (or limit deviations) are taken on the recommendations for International Tolerance Grades (IT Grades) defined in ISO 286. There are two parts of ISO 286:

- Part 1: Basis of tolerances, deviations, and fits (*ISO 286-1:2010*)
- Part 2: Standard tolerance grades and limit deviations for holes & shafts (*ISO 286-2:2010*)

<u>Note:</u> ISO 286 mostly uses the term "**Standard Tolerance Grades**" in place of "**International Tolerance Grades**", however, these are same (refer 3.2.8.2 of ISO 286-1:2010).

International Tolerance Grade (IT Grade) provides guidance for manufacturing process capability to have a fair expectation of precision and therefore, tolerance. Below are the highlights:

- It is represented by a number IT1 to IT18. Lower the number, less is tolerance, higher is precision.
- Measuring tools need very high precision. For them, 2 more IT Grades are added, IT01 and IT0.

A2 DETERMINING TOLERANCE USING IT GRADES

IT grade can be used to determine tolerance depending on the type of product, manufacturing process and dimension of the product. Below are the steps to use IT Grade for tolerance determination.

Step 1: Determine manufacturing output category and map to a range of IT Grades as shown below in figure A.1.

IT Grade	IT1	IT2	IT3	IT4	IT5	IT6	IT7	IT8	IT9	IT10	IT11	IT12	IT13	IT14	IT15	IT16	IT17	IT18
Output Category	For Measuring Tools							For Materials										
						For Fits							For Large Manufacturing					

Figure A.1– Mapping of IT grade and output category

Step 2: Determine manufacturing process to further reduce to a range of IT Grades, as shown below in figure A.2.

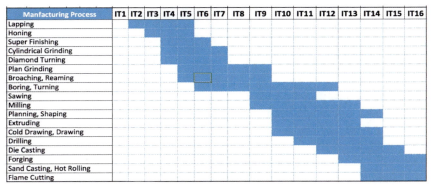

Figure A.2 – Mapping of IT grade and manufacturing process

Step 3: Refer to the figure A.3, table below to determine tolerance options for a range of IT Grades and select a tolerance satisfying your functional needs.

Above	Upto & including	IT01	IT0	IT1	IT2	IT3	IT4	IT5	IT6	IT7	IT8	IT9	IT10	IT11	IT12	IT13	IT14	IT15	IT16	IT17	IT18
		Values in μm (micron)													Values in mm (mili-meter)						
-	3	0.3	1	0.8	1.2	2	3	4	6	10	14	25	40	60	0.1	0.14	0.25	0.4	0.6	1	1.4
3	6	0.4	1	1	1.5	2.5	4	5	8	12	18	30	48	75	0.12	0.18	0.3	0.48	0.75	1.2	1.8
6	10	0.4	1	1	1.5	2.5	4	6	9	15	22	36	58	90	0.15	0.22	0.36	0.58	0.9	1.5	2.2
10	18	0.5	1	1.2	2	3	5	8	11	18	27	43	70	110	0.18	0.27	0.43	0.7	1.1	1.8	2.7
18	30	0.6	1	1.5	2.5	4	6	9	13	21	33	52	84	130	0.21	0.33	0.52	0.84	1.3	2.1	3.3
30	50	0.6	1	1.5	2.5	4	7	11	16	25	39	62	100	160	0.25	0.39	0.62	1	1.6	2.5	3.9
50	80	0.8	1	2	3	5	8	13	19	30	46	74	120	190	0.3	0.46	0.74	1.2	1.9	3	4.6
80	120	1	2	2.5	4	6	10	15	22	35	54	87	140	220	0.35	0.54	0.87	1.4	2.2	3.5	5.4
120	180	1.2	2	3.5	5	8	12	18	25	40	63	100	160	250	0.4	0.63	1	1.6	2.5	4	6.3
180	250	2	3	4.5	7	10	14	20	29	46	72	115	185	290	0.46	0.72	1.15	1.85	2.9	4.6	7.2
250	315	2.5	4	6	8	12	16	23	32	52	81	130	210	320	0.52	0.81	1.3	2.1	3.2	5.2	8.1
315	400	3	5	7	9	13	18	25	36	57	89	140	230	360	0.57	0.89	1.4	2.3	3.6	5.7	8.9
400	500	4	6	8	10	15	20	27	40	63	97	155	250	400	0.63	0.97	1.55	2.5	4	6.3	9.7
500	630			9	11	16	22	32	44	70	110	175	280	440	0.7	1.1	1.75	2.8	4.4	7	11
630	800			10	13	18	25	36	50	80	125	200	320	500	0.8	1.25	2	3.2	5	8	12.5
800	1000			11	15	21	28	40	56	90	140	230	360	560	0.9	1.4	2.3	3.6	5.6	9	14
1000	1250			13	18	24	33	47	66	105	165	260	420	660	1.05	1.65	2.6	4.2	6.6	10.5	16.5
1250	1600			15	21	29	39	55	78	125	195	310	500	780	1.25	1.95	3.1	5	7.8	12.5	19.5
1600	2000			18	25	35	46	65	92	150	230	370	600	920	1.5	2.3	3.7	6	9.2	15	23
2000	2500			22	30	41	55	78	110	175	280	440	700	1100	1.75	2.8	4.4	7	11	17.5	28
2500	3150			26	36	50	68	96	135	210	330	540	860	1350	2.1	3.3	5.4	8.6	13.5	21	33

Nominal Size in mm — International Tolerance Grades (IT Grades) — Standard Tolerance Values

Figure A.3 – Mapping of IT grade and output category

143

Note 1: **All values provided in above table is based on following formula:**

$$T = 10^{\,0.2 \times (ITG-1)} \times (\,0.45 \times D^{-1/3} + 0.001 \times D), \text{ where}$$

T is tolerance in micrometer [μm] *– to be calculated*

D is geometric mean dimensions in millimeters [mm] *– for 100 mm dia.,*
range 80-120, D = (80x120)^{-1/2}

ITG is IT Grade, a positive integer *– E.g. ITG = 6 in IT6*

Note 2: **IT Grades can be extended beyond IT18.** Simple multiply "tolerance at the fifth place earlier" by 10. For example, IT19 column will be IT14x10, IT20 column will be IT15x10, and so on.

A3 DETERMINING FUNDAMENTAL DEVIATION (USING TOLERANCE CLASS)

After learning about finding tolerances, we are going to learn about finding fundamental tolerance. Together these will provide all ***linear*** dimensions and tolerances details for manufacturing. Remember, we still have to provide geometric dimensions and tolerances for features.

A4 TOLERANCE CLASS

ISO 286 defines "*Tolerance class*" to provide different options for fundamental deviation. It has two parts:

1. **Letter(s)** - Upper-case (A to ZC) for holes and lowercase (a to z) for shafts, excluding I/i; L/l; O/o; Q/q; W/w, to avoid confusion between letter I and number 1, letter O and number 0, letter Q and number 0, letter q and number 9, and between letter w and letter v.

2. **Number** – Representing the IT grade.
   For Example, H7 represents a hole with IT Grade 7, h6 represents a shaft with IT Grade 6

Figure A.4, given below, provides tolerance class graph for different types of fits.

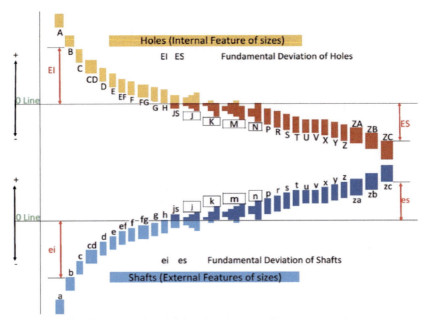

Figure A.4 – Tolerance class graph and fits

Selection of tolerance class: Among such a wide range of tolerance classes, ISO 286-1:2010 suggests preferred combinations as given in figure A.5 for shafts and A.6 for holes.

Hole Basis	Tolerance Classes for Shafts																
	Clearance Fits							Transition Fits				Interference Fits					
H 6						g5	h5	js5	k5	m5		n5	p5					
H 7					f6	g6	h6	js6	k6	m6	n6		p6	r6	s6	t6	u6	x6
H 8				e7	f7		h7	js7	k7	m7					s7		u7	
H 8			d8	e8	f8		h8											
H 9			d6	e8	f8		h8											
H 10	b9	c9	d9	e9			h9											
H 11	b11	c11	d10				h10											

Figure A.5 – Tolerance class for shafts

Shaft Basis	Tolerance Classes for Holes															
							Transition Fits				Interference Fits					
H 6					g5	H6	JS6	K6	M6		N6	P6				
H 7				F7	g6	H7	JS7	K7	M7	N7		P7	R7	S7	T7	U7	X7
H 8			E8	F8			H8									
		D9	E9	F9			H9									
H 9			E8	F8			H8									
H 10		D9	E9	F9			H9									
H 11	B11	C10	D10				H10									

Figure A.6 – Tolerance class for holes

A5 MOST WIDELY USED TOLERANCE CLASS COMBINATIONS

Figure A.7, given below provides few most widely used tolerance class combinations in the industry.

	ISO Symbol		Description	
	Hole Basis	Shaft Basis		
Clearance Fits	H11/c11	C11/h11	*Loose Running Fits:* For wide commercial tolerances	More Clearance ---->
	H9/d9	D9/h9	*Free Running Fits:* Not for high accuracy but good for -High temperature variation -High running speed -Heavy pressure	
	H8/f7	F8/h7	*Close Running Fits:* Good for accuracy at moderate speed and pressure	
	H7/g6	G7/h6	*Sliding Fits:* Not for free running, but for turn freely and accuracy	
Transition Fits	H7/h6	H7/h6	*Locational Clearance Fits;* Good for locating parts with easy assemble and de-assemble	
	H7/k6	K7/h6	*Locational Transition Fits:* for accurate location, compromise between clearance and interference.	
	H7/n6	N7/h6	*Locational Transition Fits:* for more accurate location, greater interference.	More Interference <----
Interference Fits	H7/p6	P7/h6	*Locational Interference Fits:* for rigidity nd alignment, with accuracy, without special bore pressure	
	H7/s6	S7/h6	*Medium Drive Fits:* for ordinary steel parts or shrink fits on light section; highest fit useable with cast iron	
	H7/u6	U7/h6	*Force Fits:* Suitable for highly stressed parts, for shrink fits at very high pressure	

Figure A.7 – Most widely used tolerance class combinations

ISO 286-2:2010 provides upper and lower limit deviations for a combination of nominal dimension and tolerance class. Let's find out upper and lower limit deviation for a 60 mm shaft for tolerance class M6. We will refer to table 9 of ISO 286-2:2010, part of which is given below in figure A.8.

Nominal size mm		M								N								
Above	Up to and including	3	4	5	6	7	8	9	10	3	4	5	6	7	8	9a	10a	11a
–	3a	-2 / -4	-2 / -5	-2 / -6	-2 / -8	-2 / -12	-2 / -16	-2 / -27	-2 / -42	-4 / -6	-4 / -7	-4 / -8	-4 / -10	-4 / -14	-4 / -18	-4 / -29	-4 / -44	-4 / -64
3	6	-3 / -5,5	-2,5 / -6,5	-3 / -8	-1 / -9	0 / -12	+2 / -16	-4 / -34	-4 / -52	-7 / -9,5	-6,5 / -10,5	-7 / -12	-5 / -13	-4 / -16	-2 / -20	0 / -30	0 / -48	0 / -75
6	10	-5 / -7,5	-4,5 / -8,5	-4 / -10	-3 / -12	0 / -15	+1 / -21	-6 / -42	-6 / -64	-9 / -11,5	-8,5 / -12,5	-8 / -14	-7 / -16	-4 / -19	-3 / -25	0 / -36	0 / -58	0 / -90
10	18	-6 / -9	-5 / -10	-4 / -12	-4 / -15	0 / -18	+2 / -25	-7 / -50	-7 / -77	-11 / -14	-10 / -15	-9 / -17	-9 / -20	-5 / -23	-3 / -30	0 / -43	0 / -70	0 / -110
18	30	-6,5 / -10,5	-6 / -12	-5 / -14	-4 / -17	0 / -21	+4 / -29	-8 / -60	-8 / -92	-13,5 / -17,5	-13 / -19	-12 / -21	-11 / -24	-7 / -28	-3 / -36	0 / -52	0 / -84	0 / -130
30	50	-7,5 / -11,5	-6 / -13	-5 / -16	-4	0 / -25	+5 / -34	-9 / -71	-9 / -109	-15,5 / -19,5	-14 / -21	-13 / -24	-12 / -28	-8 / -33	-3 / -42	0 / -62	0 / -100	0 / -160
50	80			-6 / -19	-5 / -24	0 / -30	+5 / -41					-15 / -28	-14 / -33	-9 / -39	-4 / -50	0 / -74	0 / -120	0 / -190

Figure A.8 – Part of table 9 of ISO 286-2:2010

As highlighted in the table, the upper limit deviation is -5 μm and lower limit deviation is -24 μm. It means the hole must be made between size 59.976 and 59.995 mm.

This seems to be very high precision to maintain. Isn't it? This is the reason; it is not advised by ISO. Verify it in the chart given above. At this point, the designer has to take a call, either to provide a bigger allowance for easy manufacturing, if it meets intended functionality or to ask for it, even at the higher cost of manufacturing, as per need. This decision is the most crucial part of a designer.

QUESTION BANK

The approach of the question bank

Generally, we find traditional books asking questions at the end of every chapter and also provide some exercises to expertise the topic of the chapter. This is a well experienced streamlined approach. It is good for sequential reading and academics. In practical situations, high-grade tests or during job interviews, the situation differs. Questions do come randomly. You are forced to recall many areas of studies and your experiences to answer these questions. This section is intended to thoroughly test your current level of understanding of GD&T.

Following areas are covered in two different sections:
Q1. Fundamental based questions
Q2. Design calculation based questions

You should attempt all the questions. If answers do not match, then investigate deeper and make sure you understand the reason behind the correct answers and correct your understanding.

SELF EVALUATION AND NEXT STEPS

Below is the advisory for the next step, based on your performance after attempting these questions

Your Performance	Next step
0-30%	Read again from first chapter. Probably required level of attention was missing. This time get it.
31-60%	Read from 4th chapter. Focus on examples.
61-80%	You can claim to understand GD&T, little more practice will help.
81-99%	Bravo!. You were purpose of my writing this book. Enjoy, have fun, and move ahead to claim the GD&T expert position. Well done!
100%	You are guru! Who am I to advice you?

All the Best!

Q1. FUNDAMENTAL BASED QUESTIONS

1. What are the main purposes of GD&T concept?
   A. Saving cost
   B. Indicating tolerances for different geometric features
   C. Communication of functional requirement of the features of the parts or between parts of the assemblies
   D. All of the above

2. GD&T provided tight or loose tolerances?
   A. Intentional tight tolerances for better quality
   B. Derivable loose tolerances for easy manufacturing
   C. It depends on the features on which GD&T is applied
   D. GD&T do not change the values of any tolerances

3. Why GD&T is called a **functional language** between the designer and the manufacturers?
   A. It works as a mathematical function to calculate tolerances.
   B. It works on the concepts of geometric functions.
   C. Designer provides functionality requirements of the parts or between parts of the assemblies which are derived from the design requirements.
   D. It is not a functional language, it is a tolerancing language.

4. How can GD&T reduce the cost of the manufacturing process?
   A. Communication of the functional design requirements, manufacturer can derive additional bonus tolerances to reduce the chance of rejection.
   B. Cost remains unchanged, because GD&T do not talk about materials to be used or machining processes to be followed.
   C. In fact, it will increase the cost of the manufacturing due to additional GD&T requirements.
   D. Cost may increase or decrease, depending on the types of the GD&T requirement required by the designer.

5. What is a Basic Dimension?
   A. It is the dimension of a feature on which any tolerance is not permissible and manufacturer has to produce at exact size.
   B. It is the dimension which is presented in basic unit.
   C. It is dimension of features at the base of the equipment.
   D. It is theoretical exact size on which tolerances are given.

6. What is the condition of an internal feature when it measures the largest size within design limits or weighs the least
   A. MMC
   B. LMC
   C. RFS
   D. None of the above

7. How reference only dimensions are mentioned?
   A. With bracket
   B. With word REF
   C. Any of the above
   D. None of the above

8. How the dimensions without tolerances are written on drawing?
   A. With sign ±0.00
   B. Written inside a box
   C. Not allowed
   D. Written inside brackets

9. What is a feature is mechanical engineering?
   A. It is any physical portion (curved or flat) of a part
   B. It is important portion of a part which interacts with other parts in any assembly.
   C. It is any portion of the part which is exposed to be visible.
   D. It is main feature of ant machine.

10. What are feature of size (FOS)?
    A. These are features without any tolerance.
    B. These are the features which are given a dimension with GD&T tolerances.
    C. These are features which has opposite physical points which can be measured and also it can be used for references.
    D. It is geometric shape of any feature, e.g., circular or straight.

11. Elliptical hole can be considered as a feature of size?
    A. Yes, internal size can be measured between opposed end.
    B. No, it does not have consistent size between any two opposed points.

12. Upper limit + Lower limit = Total tolerance
    A. True
    B. False

13. If upper deviation = lower deviation then it will be basic size.
    A. True
    B. False

14. MMC represents maximum size of any feature.
    A. True
    B. False

15. Which statement is true?
    A. Upper deviations, lower deviations, and fundamental deviations are unwanted deviation.
    B. Upper deviations, lower deviations, and fundamental deviations are wanted deviation.
    C. Upper deviations, lower deviation are unwanted deviations but fundamental deviations are wanted deviations.
    D. Upper deviations, lower deviation are wanted deviations but fundamental deviations are unwanted deviations.

16. Which is correct statement for Regardless of feature size (RFS) ?
    A. Regardless of position of the feature, dimension tolerance must be met.
    B. Regardless of the manufacturing process, geometric tolerance must be met.
    C. Regardless of datum condition of the feature control, geometric tolerance must be met.
    D. Regardless of the actual size of the feature, geometric tolerance must be met.

17. In case of RFS control, bonus tolerance is always zero.
    A. True   B. False

18. For any systems of fits (hole and shaft), maximum clearance is
    A. MMC shaft – MMC hole   B. LMC shaft – LMC hole
    C. MMC hole – MMC shaft   D. LMC hole – LMC shaft

19. For any systems of fits (hole and shaft), minimum interference is
    A. MMC shaft – MMC hole   B. LMC shaft – LMC hole
    C. MMC hole – MMC shaft   D. LMC hole – LMC shaft
20. To design a clearance fit, which kind of analysis you need to perform in design process?
    A. MMC for both hole and shaft
    B. LMC for nth hole and shaft
    C. MMC for hole andLMC for shaft
    D. LMC for hole and MMC for shaft

21. What is difference between a Datum and Datum Feature?
    A. Datum is universal, Datum feature is local to a feature
    B. Datum is hypothetical, Datum feature is actual
    C. Datum is theoretical, Datum feature is practical
    D. Both are same.

22. What is relationship between Primary Datum and Degree of freedom (DOF)?
    A. Primary datum controls one DOF
    B. Primary datum controls two DOF
    C. Primary datum controls three DOF
    D. Primary datum controls all six DOF

23. Is it necessary for primary, secondary, and tertiary datums to be mutually perpendicular to each other?
    A. Yes, as it is taken in datum reference frame
    B. No. In fact datum need not be a plane. It can even be a curved profile

24. As a designer, we provide functional or manufacturing datum?
    A. Functional datum   B. Manufacturing datum

25. It is necessary to restrict all six DOF for manufacturing?
    A. Yes, that is why three datums are used.
    B. No, take example of turning operation in which one rotational degree of freedom is unrestricted.

26. When datum target is used?
    A. When datum is decided based on target of manufacturing.
    B. When datum feature is too large or uneven.
    C. When datum is based on target of finished feature.
    D. It is extremely difficult to use. So practically it is never used.

27. Circularity is applicable to independent or assembly feature?
    A. Independent feature
    B. Assembly feature
    C. Both of the above
    D. It depends upon type of manufacturing process.

28. Do we require datum for circularity?
    A. Yes   B. No
    C. May be (optional)   D. Datum is not relevant to circularity

29. Do we need datum for profile of a line?
    A. Yes
    B. No
    C. May be (optional)
    D. Datum is not applicable for profile of a line

30. In which GD&T tolerance, datum is mandatory?
    A. Position   B. Straightness
    C. Flatness   D. Profile of a surface

31. Feature control frame is
    A. It is GD&T representation in a box to communicate manufacturing process and geometric tolerances.
    B. It is a method to communicate geometric controls, by means of defined tolerances, represented in a box structure.
    C. It is part control details for quality inspector.
    D. It is a physical clamp like frame on which part is fixed to control any feature of a part during manufacturing process.

32. We can apply material condition on feature as well as datum.
    A. True   B. False

33. Can we provide GD&T without dimension and dimensional tolerance?
    A. Yes, GD&T is independent of dimensional tolerance
    B. No, GD&T is applied on top of dimensional tolerance

34. Tolerance zone for straight is
    A. Rectangular plane for straightness on a surface
    B. Cylindrical volume for straightness of an axis
    C. Both of the above
    D. None of the above

35. Is material condition (MMC/LMC) mandatory for straightness?
    A. Yes, for both of axis and surface straightness.
    B. No, for none of axis and surface straightness.
    C. Yes, only for axis straightness.
    D. Yes, only fro surface straightness.

36. Is it possible to have axis straightness greater than position tolerance?
    A. Yes   B. No

37. If a dimension is whole number, say 20 mm, then what is the right way to show it according to ASME standard.
    A. 20   B. 20.0   C. 20.00   D. 20.000

38. According to ASME Y14.5-2009, per unit tolerance concept is applicable only to
    A. Straightness   B. Flatness
    C. Both of the above   D. None of the above

39. Datum is mandatory for circularity.
    A. True   B. False

40. Material condition (MMC/LMC) is mandatory for circularity.
    A. True   B. False

41. Cylindricity is inclusive of:
    A. Straightness   B. Circularity
    C. Runout   D. Only A and B

42. Material condition is optional for flatness tolerance, because
    A. it depends on fixture used during manufacturing process
    B. for independent flat feature datum is not needed but having another interacting feature, datum may be needed.
    C. flatness can be controlled by parallelism
    D. it is basically multiple straightness for which material condition is not needed
    E. it gives same output with or without material condition

43. Profile of a line is controlled with material condition.
    A. True   B. False

44. Profile of a line is always 2D in nature.
    A. True   B. False

45. If we consider 2D profile of a surface then it is same as
    A. Profile of a line   B. Flatness
    C. Parallelism   D. None of the above

46. We can control flatness with surface profile but we don't do so. Why?
    A. Flatness is 2D however profile of a surface is 3D
    B. Controlling profile of a surface is complicated than controlling flatness
    C. Using parallelism is better than profile of a surface
    D. Above statement is incorrect.

47. In case of angularity, sometime we use only one datum and sometimes we use 2 datums. How to decide how may datum to be used?
    A. It depends of functionality, no fixed rule can be derived
    B. For angularity between planes, we use only one datum, but for angularity between plane and a line then two datums becomes mandatory to define exact position of the line.
    C. We may need to use up to three datums for angularity.
    D. Datum is optional for angularity.

48. Perpendicularity is inclusive of:
    A. Straightness
    B. Flatness
    C. Position
    D. Only A and B

49. Parallelism is always used with one datum. Two datums are impossible to use in case of parallelism.
    A. True
    B. False

50. Material condition is _____ for position tolerance.
    A. Required
    B. Not required
    C. Optional
    D. No fixed rule is defined

51. Bonus tolerance is independent of material condition.
    A. True
    B. False

52. Why concentricity is rarely used in the industry?
    A. Functionality of concentricity can easily be achieved through position and straightness.
    B. Concentricity is measured by center of mass distribution, which is extremely difficult to use.
    C. Position of concentric features are difficult to reach for measurement of concentricity.
    D. There is no equipment to check concentricity.

53. Datum is mandatory for concentricity
    A. True
    B. False

54. Material condition is mandatory for concentricity.
    A. True
    B. False

55. Symmetry includes:
    A. Flatness
    B. Parallelism
    C. Position
    D. All of the above

56. What is difference between Runout and Total runout?
    A. Runout is used for one part, total runout is used for multiple parts, all at a time
    B. Runout is used at a cross section of a shaft, total runout is used for a length of a shaft
    C. Total runout can be used for inclined cylindrical profile but runout cannot be used for same.
    D. Runout works with one datum but total runout works with at least two datums.

57. Datum axis derived for runout or total runout can be derived with help if two cylindrical datum features, say A and B, can be represented as:
   A. A+B
   B. A–B
   C. A & B
   D. A to B

58. Runout is inclusive of
   A. Position, and straightness
   B. Position, and circularity
   C. Straightness, and circularity
   D. Position, straightness, and circularity

59. Total runout is inclusive of
   A. Concentricity, and perpendicularity
   B. Cylindricity, and circularity
   C. Straightness and runout
   D. All of the above

60. Material condition is required for runout and total runout
   A. True
   B. False

61. Which of the following are types of feature control(s) when we try to control any feature with than one geometric tolerances?
   A. Multiple feature control
   B. Composite feature control
   C. Combines feature control
   D. All of the above

62. What is a boundary in terms of dimensional tolerance (DT), geometric tolerance (GT), and bonus tolerance (BT)
   A. It is maximum deviation due to DT
   B. It is combined maximum deviation due to DT + GT
   C. It is combined maximum deviation due to GT + BT
   D. It is combined maximum deviation due to DT + GT + BT

63. Inner boundary of a hole is same as:
   A. Size at Maximum material condition (MMC)
   B. Size at Least material condition (LMC)
   C. Size at Maximum material boundary (MMB)
   D. Size at Least material Boundary (LMB)

64. Inner boundary of a shaft is same as:
    A. Size at Maximum material condition (MMC)
    B. Size at Least material condition (LMC)
    C. Size at Maximum material boundary (MMB)
    D. Size at Least material Boundary (LMB)

65. Outer boundary of a hole is same as:
    A. Size at Maximum material condition (MMC)
    B. Size at Least material condition (LMC)
    C. Size at Maximum material boundary (MMB)
    D. Size at Least material Boundary (LMB)

66. Outer boundary of a shaft is same as:
    A. Size at Maximum material condition (MMC)
    B. Size at Least material condition (LMC)
    C. Size at Maximum material boundary (MMB)
    D. Size at Least material Boundary (LMB)

67. MMB will be same as actual MMB when location or orientation tolerance for datum feature is
    A. Zero at MMC
    B. Zero at LMC
    C. Zero at MMC and LMC
    D. Zero at RFS

68. LMB will be same as actual LMB when location or orientation tolerance for datum feature is
    A. Zero at MMC
    B. Zero at LMC
    C. Zero at MMC and LMC
    D. Zero at RFS

69. Virtual condition included bonus tolerance
    A. True   B. False

70. Resultant condition included bonus tolerance
    A. True   B. False

71. Actual mating envelop is closet fit counterpart
    A. used in the actual assembly
    B. with contact at all points
    C. with no contact but closet to actual profile
    D. with maximum contacts being perfect in shape

72. Related actual mating envelope is an actual mating envelop
    A. related to assembly with counterpart
    B. found when considering location conditions by datum
    C. found when considering orientation conditions by datum
    D. All of the above

73. Unrelated actual mating will always be greater than related actual mating
    A. True   B. False

74. _____ and _____ of features are controlled by Rule #1
    A. Size and Tolerance
    B. Size and form
    C. Form and envelop
    D. Envelop and boundary

75. Only Individual features are controlled by Rule # 1
    A. True   B. False

76. Rule # 2 makes one of the following to be mandatory, if applicable, to be mentioned in feature control frame.
    A. MMC, LMC, RFS   B. MMC, LMC
    C. MMC   D. Nothing is mandatory

77. Translation modifier is used
    A. to translate the GD&T details to another language
    B. when MMC or LMC can be changed
    C. when dimensional tolerance can be used as form tolerance
    D. when datum translation is allowed for better functionality

78. Projected tolerance zone may be used for
    A. Screw   B. Stud
    C. Pin   D. All of the above

79. If length of projected tolerance zone is shorter then effective projected tolerance zone is
    A. Shorter
    B. Larger
    C. Remains same
    D. Depends on material condition <MMC/LMC

80. Free state is a type of modifier which defines geometric tolerance
    A. of rigid parts
    B. of non-rigid parts
    C. at normal room temperature
    D. at working condition temperature

81. Tangent plane modifier is
    A. a method to measure the profile of a feature by using only higher points of the feature
    B. used in place of flatness or parallelism when another component needs to be mounted on the feature
    C. measured by another plane placed on top of the profile
    D. All of the above

82. Unequally disposed profile modifier controls
    A. geometric tolerances of disposable parts
    B. unequal size of same profile to dispose/reject bad parts
    C. the distribution of tolerance between upper and lower limits
    D. the resultant profile to have equal upper and lower deviation

83. In terms of unequally disposed modifier, upper deviation is called _____ and lower deviation is called _____.
    A. minimum addition, maximum removal
    B. minimum addition, minimum removal
    C. maximum addition, maximum removal
    D. maximum addition, minimum removal

84. Independency modifier is used
    A. to make any datum feature to work independent of any other datum of combination of datums
    B. to make any feature control independent of any datum
    C. to independently provide geometric tolerance without considering actual functionality
    D. for mass production of any independent part

85. Below are the types of stack tolerance analysis
    A. Worst case analysis
    B. Cumulative sum analysis
    C. Statistical tolerance analysis
    D. Only A and C

86. Normal distribution statistical analysis assumes
    A. tolerance of an assembly is average of tolerances of all components
    B. tolerance of an assembly is arithmetic mean of tolerances of all components
    C. tolerance of an assembly is square root of sum of square of tolerances of all components
    D. tolerance of an assembly is maximum tolerances of all components

87. Statistical tolerance modifier relaxes the tolerance requirements, but it should be used only at a place
    A. where statistics experts are available
    B. where statistical procedure controls are used
    C. where statistical calculations are done on CAD software
    D. where six sigma certification is available

88. Continuous feature modifier is used for a surface
    A. which has no breaks between two ends
    B. which has repeated surface of same specifications
    C. which has continuously changing profile
    D. None of the above

89. Controlled radius modifier is used
    A. when smooth curve is required
    B. when feature is complete circular in shape
    C. for continuously rotating circular features
    D. when radius is controlled by designers

90. Dimension origin modifier is used
    A. for unsymmetrical parts
    B. to denote the starting point of measurement
    C. to define reference system or plane for quality assurance
    D. All of the above

91. Between modifier is used
    A. when same geometric tolerance applies to all features falling between two given points.
    B. when geometric tolerance is given as a range and actual value is expected to remain within the given range
    C. Both of the above
    D. None of the above

92. All around modifier zone is 2D and All over modifier zone is 3D.
    A. True
    B. False

93. Individually modifier is generally used
    A. for repeated pattern of feature(s)
    B. for large work piece and measurement all dimensions becomes difficult
    C. when multiple features need to function as a group.
    D. All of the above

94. ISO 286-1:2010 refers IT grade as:
    A. International tolerance grade
    B. Intentional tolerance grade
    C. Standard tolerance grade
    D. None of the above

95. In IT grade always remains between IT1 to IT18
    A. True   B. False

96. In IT grade, less is the number:
    A. Higher is precision
    B. Less is tolerance
    C. Costlier is manufacturing
    D. All of the above

97. Determining IT grade will help in deciding geometric tolerances
    A. True   B. False

98. Tolerance class, as defined in ISO-286, is used to decide:
    A. Dimensional tolerance
    B. Geometric tolerance
    C. Fundamental deviation
    D. IT Grade

99. In tolerance class H7, H represent _____ and number '7' represents
    _____
    A. 'Hole' and 'IT Grade'
    B. 'Hole' and 'Size of feature in mm'
    C. 'Shaft' and 'IT Grade'
    D. 'Shaft' and 'Size of feature in mm'

100. IT grades used for highest precision measuring instruments are:
    A. IT01
    B. IT0
    C. Both of the above
    D. None of the above

Q2. DESIGN CALCULATION BASED QUESTION

You are designing a clearance fit of basic size = 10.00. You kept upper deviation of the shaft as 0.10, lower deviation of the shaft as 0.20, upper deviation of hole as 0.25, and lower deviation of the hole as 0.15. Calculate following values:

1. MMC of the shaft
    A. 10.00   B. 9.90   C. 9.80   D. 10.10

2. LMC of the shaft
    A. 10.00   B. 9.90   C. 9.80   D. 10.10

3. Total dimensional tolerance of the shaft
    A. 0.00   B. 0.10   C. 0.20   D. 0.25

4. Fundamental deviation of the shaft
    A. 0.00   B. 0.10   C. 0.20   D. 0.25

5. MMC of the hole
    A. 10.00   B. 10.15   C. 10.25   D. 10.40

6. LMC of the hole
    A. 10.00   B. 10.15   C. 10.25   D. 10.40

7. Total dimensional tolerance of the hole
    A. 0.00   B. 0.10   C. 0.15   D. 0.20

8. Fundamental deviation of the Hole
    A. 0.00   B. 0.10   C. 0.15   D. 0.20

9. Maximum gap of the clearance fit
    A. 0.25   B. 0.35   C. 0.45   D. 0.55

10. Minimum gap of the clearance fit
    A. 0.25   B. 0.35   C. 0.45   D. 0.55

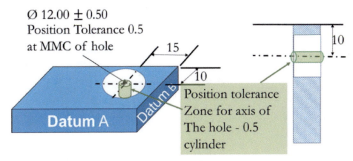

Ø 12.00 ± 0.50
Position Tolerance 0.5
at MMC of hole

Figure Q2.1: Situation for question 1 to 4

In the figure given above, find following values:

11. MMC size of hole
    A. 11.00   B. 11.50   C. 12.50   D. 13.50

12. MMC virtual condition size of the hole
    A. 11.00   B. 11.50   C. 12.50   D. 13.50

13. LMC size of the hole
    A. 11.00   B. 11.50   C. 12.50   D. 13.50

14. Total positional tolerance for the hole if the actual size is 12.20
    A. 0.50   B. 0.70   C. 1.00   D. 1.20

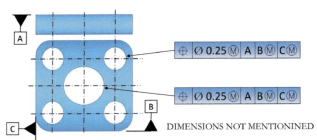

Figure Q2.1: Situation for question 5

15. When datum features B and C move away from MMC by 0.10,
    the total positional tolerance zone size of holes will be:
    A. 0.25   B. 0.30   C. 0.35   D. 0.45

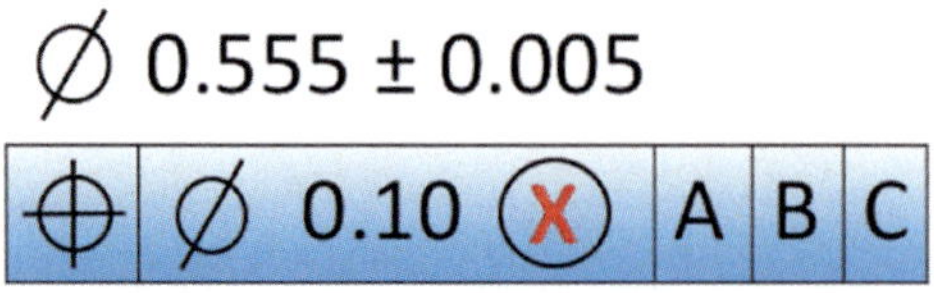

⌀ 0.555 ± 0.005

⊕	⌀ 0.10 Ⓧ	A	B	C

Figure Q2.2: Details of a hole

In figure Q2.2, find total positional tolerance:

16. if actual size of hole is 0.555 and X = M
    A. 0.10   B. 0..105   C. 0.102   D. 0.110

17. if actual size of hole is 0.550 and X = M
    A. 0.10   B. 0.105   C. 0.102   D. 0.110

18. if actual size of hole is 0.560 and X = M
    A. 0.10   B. 0.105   C. 0.102   D. 0.110

19. if actual size of hole is 0.552 and X = M
    A. 0.10   B. 0.105   C. 0.102   D. 0.110

20. if actual size of hole is 0.555 and X = L
    A. 0.10   B. 0.105   C. 0.095   D. 0.110

21. if actual size of hole is 0.550 and X = L
    A. 0.10   B. 0..105   C. 0.095   D. 0.110

22. if actual size of hole is 0.560 and X = L
    A. 0.10   B. 0.105   C. 0.095   D. 0.110

23. if actual size of hole is 0.552 and X = L
    A. 0.10   B. 0.102   C. 0.108   D. 0.110

24. if actual size of hole is 0.555 and X is not mentioned, i.e., RFS
    A. 0.10   B. 0.105   C. 0.095   D. 0.110

25. if actual size of hole is 0.560 and X is not mentioned, i.e., RFS
    A. 0.10   B. 0.105   C. 0.095   D. 0.110

A **hole** has permissible diameter between 30.0 and 31.0. The positional tolerance is 0.10 @ **MMC**. What will be the values of:

26. Maximum bonus tolerance
    A. 0.00   B. 1.00   C. 0.50   D. 0.01

27. Virtual condition boundary
    A. 29.90   B. 30.00   C. 30.10   D. 31.10

28. Resultant condition boundary
    A. 31.10   B. 31.11   C. 32.10   D. 32.11

A **hole** has permissible diameter between 30.0 and 31.0. The positional tolerance is 0.10 @ **LMC**. What will be the values of:

29. Virtual condition boundary
    A. 28.90   B. 30.00   C. 30.10   D. 31.10

30. Resultant condition boundary
    A. 28.90   B. 31.11   C. 32.10   D. 32.11

A **shaft** has permissible diameter between 30.0 and 31.0. The positional tolerance is 0.10 @ **MMC**. What will be the values of:

31. Virtual condition boundary
    A. 28.90   B. 30.00   C. 30.10   D. 31.10

32. Resultant condition boundary
    A. 28.90   B. 31.11   C. 32.10   D. 32.11

A **shaft** has permissible diameter between 30.0 and 31.0. The positional tolerance is 0.10 @ **LMC**. What will be the values of:

33. Virtual condition boundary
    A. 28.90   B. 29.90   C. 30.10   D. 31.90

34. Resultant condition boundary
    A. 28.90   B. 31.11   C. 32.10   D. 32.11

A **hole** has permissible diameter between 30.0 and 31.0. The positional tolerance is 0.10 @ **RFS**. What will be the sizes of:

35. Maximum bonus tolerance
    A. 0.00   B. 1.00   C. 0.50   D. 0.01

36. Inner boundary
    A. 28.90   B. 29.90   C. 30.10   D. 31.90

37. Resultant condition boundary
    A. 28.90   B. 31.10   C. 32.10   D. 32.11

A **shaft** has permissible diameter between 30.0 and 31.0. The positional tolerance is 0.10 @ **RFS**. What will be the sizes of:

38. Maximum bonus tolerance
    A. 0.00   B. 1.00   C. 0.50   D. 0.01

39. Inner boundary
    A. 28.90   B. 29.90   C. 30.10   D. 31.90

40. Resultant condition boundary
    A. 28.90   B. 31.10   C. 32.10   D. 32.11

You are designing an interference fit of basic size = 10.00. You kept shaft dimension as 10.10-10.15 and hole dimension as 9.95-9.90 such that it an interference fit in all condition. You are designing to put geometric tolerances. Answer following design questions:

41. If no geometric tolerance is applied on either of the features, what is the minimum interference designed in above case?
    A. 0.00   B. 0.05   C. 0.10   D. 0.20

42. If geometric tolerance is applied on the shaft then what is maximum bonus tolerance you are allowing to manufacturer?
    A. 0.00   B. 0.05   C. 0.10   D. 0.20

43. If geometric tolerance is applied on the hole then what is maximum bonus tolerance you are allowing to manufacturer?
    A. 0.00   B. 0.05   C. 0.10   D. 0.20

44. If no geometric tolerance is applied on hole, what can be maximum available value of position tolerance on shaft at LMC such that minimum interference of 0.10 is available?
    A. 0.00   B. 0.05   C. 0.10   D. 0.20

45. If no geometric tolerance is applied on shaft, what can be maximum available value of position tolerance on hole at LMC such that minimum interference of 0.15 is available? 0.00
    A. 0.00   B. 0.05   C. 0.10   D. 0.20

46. For greater than 0.00 interference, combined total geometric tolerance available together for both the features will be:
    A. 0.00   B. 0.05   C. 0.10   D. 0.20

| ◠ | 3.0 Ⓤ 1.0 | A | B | C |

47. In GD&T notation given above, the value of lowed deviation is:
    A. 0.0   B. 1.0   C. 2.0   D. 3.0

48. A shaft has size 10±0.01 and axis straightness Ø 0.001 @ MMC. If actual size is 10.005 then diameter of axis tolerance zone is:
    A. 0.001   B. 0.005   C. 0.006   D. 0.011

49. A table top has dimension of 20±0.02. What can be maximum value of flatness if only applicable to upper face of the table top?
    A. 0.00   B. 0.01   C. 0.02   D. 0.03

50. Consider a clearance fit design with basic size 15.00 mm, 0.1% fundamental deviation and IT7 grade fit. Find MMC diameter of hole and MMC diameter of shaft.
    A. MMC of hole: 20.000; MMC of shaft: 20.000
    B. MMC of hole: 20.020; MMC of shaft: 19.980
    C. MMC of hole: 20.041; MMC of shaft: 19.980
    D. MMC of hole: 20.241; MMC of shaft: 19.959

Answers: Q1-Fundamental based questions

1	2	3	4	5	6	7	8	9	10
D	B	C	A	D	B	C	B	A	C

11	12	13	14	15	16	17	18	19	20
B	A	B	B	C	D	A	D	B	A

21	22	23	24	25	26	27	28	29	30
B	C	B	A	B	B	A	B	C	A

31	32	33	34	35	36	37	38	39	40
B	A	B	C	C	B	A	C	B	B

41	42	43	44	45	46	47	48	49	50
D	B	B	A	B	B	B	D	B	A

51	52	53	54	55	56	57	58	59	60
B	B	A	B	D	B	B	B	D	B

61	62	63	64	65	66	67	68	69	70
D	D	C	D	D	C	A	B	B	A

71	72	73	74	75	76	77	78	79	80
D	D	A	B	A	B	D	D	B	B

81	82	83	84	85	86	87	88	89	90
D	C	C	B	D	C	B	B	A	D

91	92	93	94	95	96	97	98	99	100
A	A	D	C	B	D	B	C	A	C

Answers: Q2-Design calculation based questions

1	2	3	4	5	6	7	8	9	10
B	C	B	B	B	C	B	C	C	a

11	12	13	14	15	16	17	18	19	20
B	A	C	D	C	B	A	D	C	b

21	22	23	24	25	26	27	28	29	30
D	A	C	A	A	B	A	C	D	A

31	32	33	34	35	36	37	38	39	40
D	A	B	C	A	B	B	A	B	B

41	42	43	44	45	46	47	48	49	50
D	B	B	B	A	C	C	C	C	C

CONCLUDING NOTES

Purpose of GD&T is to save money
- By making requirements clearer to the manufacturer.
- By making tolerances explicitly documented.
- By finalizing the acceptance criteria before production.

GD&T is relatively new in the mechanical industry. It is maturing and new concepts are shaping up. In this evolving phase, you should feel privileged to be a part of the journey.

Going to basics, **GD&T is to put additional geometric tolerances on top of dimensional tolerances**. There are many standards to decide dimensional tolerances but there is no globally accepted single standard yet available to derive this geometric tolerance. This is obvious for two reasons:
1. The geometric concept is in an evolving phase
2. Geometric tolerances are related to the functional requirement and there could be many combinations of functional requirements for which writing standard will be difficult, if not impossible.

Convert the challenge into an opportunity. This is where your designer hat can put you in driving seat. You just need to understand the most important functions and put those priorities in terms of GD&T language. You would become GD&T expert before you realize.

There is no perfection achieved yet. The entire world is working towards it to make it simpler. So you are allowed to make mistakes and learn. It's not only okay but it's the only way you will learn GD&T as a practitioner.

Across the globe, as we read, there is a **huge scarcity of GD&T expert**. Demand is more as word realized the potential to save money by standardizing tolerancing and manufacturing.

This is the right time to make your full-throttle move towards becoming a GD&T expert to advance in your career.

Finally, I would urge you to review this book on the platform which you got this book. It may be Amazon, iBook, etc., but you must take out some time to put your true findings of this book. It will help me to improve this book and release a new edition with more accurate and beneficial contents. It will help fellow readers to get reading content and save time.

I thank you for spending time with me on this book. You can get in touch with me through www.azukotech.com and I would revert based on best of my capabilities.

All the Best for your future endeavours!

-Ashok Kumar

ABOUT THE AUTHOR

Ashok Kumar is basically a mechanical engineer. He was All India Rank 1 in mechanical engineering during his studies, he is M. Tech. from IIT Mumbai, India. Mr Kumar has international work experience in the USA, UK, Europe, UAE and India. He has played various roles ranging from highly technical to managerial to leadership positions. He is running his own business with global client base. He has shared his wide management experiences through another book on management and the book is a bestselling book on Amazon in the USA, UK, Germany, France and India.

This book is intended to simplify and widespread the concepts of GD&T. Mr Kumar considers it as revolutionary development for the manufacturing industry to increase quality, efficiency and reduce cost. It is the need of the industry. Therefore, he invites all mechanical engineers to join this revolutionary development and participate in it.

Made in the USA
Middletown, DE
15 August 2020

15431922R00100